Ionic Liquids IIIA: Fundamentals, Progress, Challenges, and Opportunities

ACS SYMPOSIUM SERIES **901**

Ionic Liquids IIIA: Fundamentals, Progress, Challenges, and Opportunities

Properties and Structure

Robin D. Rogers, Editor
The University of Alabama

Kenneth R. Seddon, Editor
The Queen's University of Belfast

Sponsored by the ACS Divisions
Industrial and Engineering Chemistry, Inc.
Environmental Chemistry, Inc.
Petroleum Chemistry, Inc.

American Chemical Society, Washington, DC

Library of Congress Cataloging-in-Publication Data

Ionic Liquids III : fundamentals, progress, challenges, and opportunities / Robin D. Rogers, editor, Kenneth R. Seddon, editor ; sponsored by the ACS Divisions of Industrial and Engineering Chemistry, Inc., Environmental Chemistry, Inc., Petroleum Chemistry, Inc.

    p. cm.—(ACS symposium series ; 901–902) *MVM*

    Includes bibliographical references and index.

    Contents: A. Properties and structure — B. Transformations and processes.

    ISBN 0–8412–3893–6 (v. A : alk. paper) — ISBN 0–8412–3894–4 (v. B : alk. paper)

    1. Ionic structure—Congresses. 2. Ionic solutions—Congresses. 3. Solution (Chemistry)—Congresses.

    I. Title: Ionic liquids 3. II. Rogers, Robin D.. III. Seddon, Kenneth R., 1950- IV. American Chemical Society. Division of Industrial and Engineering Chemistry, Inc. V. American Chemical Society. Division of Environmental Chemistry, Inc. VI. American Chemical Society. Division of Petroleum Chemistry, Inc. VII. Series.

QD540.I58   2005
541′.34—dc22   2004063111

The paper used in this publication meets the minimum requirements of American National S tandard for Information Sciences—Permanence of Paper for Printed Library Materials, ANSI Z39.48–1984.

PRINTED IN THE UNITED STATES OF AMERICA

Foreword

The ACS Symposium Series was first published in 1974 to provide a mechanism for publishing symposia quickly in book form. The purpose of the series is to publish timely, comprehensive books developed from ACS sponsored symposia based on current scientific research. Occasionally, books are developed from symposia sponsored by other organizations when the topic is of keen interest to the chemistry audience.

Before agreeing to publish a book, the proposed table of contents is reviewed for appropriate and comprehensive coverage and for interest to the audience. Some papers may be excluded to better focus the book; others may be added to provide comprehensiveness. When appropriate, overview or introductory chapters are added. Drafts of chapters are peer-reviewed prior to final acceptance or rejection, and manuscripts are prepared in camera-ready format.

As a rule, only original research papers and original review papers are included in the volumes. Verbatim reproductions of previously published papers are not accepted.

ACS Books Department

In Memoriam

This volume is dedicated, with deepest respect, to the memory of Professor Robert Osteryoung, the grandfather of ionic liquids, who died of amyotrophic lateral sclerosis at the age of 77, in August 2004. His insight and inspiration will be greatly missed by the entire ionic liquids community.

Contents

Structure and Spectroscopy

ix

Theory and Modeling

Physical Properties

Phase Equilibria

Preface

The chapters in these two books (*Ionic Liquids IIIA: Fundamentals, Progress, Challenges, and Opportunities: Properties and Structure* and *Ionic Liquids IIIB: Fundamentals, Progress, Challenges, and Opportunities: Transformations and Processes*) are based on papers that were presented at the symposium *Ionic Liquids: Fundamentals, Progress, Challenges, and Opportunities* at the 226th American Chemical Society (ACS) National Meeting held September 7–11, 2003 in New York. This was the third and last of a trilogy of meetings, the second of which was *Ionic Liquids as Green Solvents: Progress and Prospects* at the 224th ACS National Meeting held August 18–22, 2002 in Boston, Massachusetts, which followed, by eighteen months, the first successful ionic liquids symposium *Ionic Liquids: Industrial Applications for Green Chemistry* at the San Diego, California ACS meeting in April 2001.

The success of the New York meeting can be judged by the simple fact that we need to publish two books to include the key papers presented, in a year in which we anticipate 1000 papers concerning ionic liquids will be published. The talks showed the depth of research currently being undertaken; the broad and diverse base for activities; and the excitement and potential opportunities that exist, and are continuing to emerge, in the field. The new industrial applications raise an especial level of interest and excitement.

The New York meeting comprised ten half-day sessions that broadly reflected the areas of development and interest in ionic liquids. We are indebted to the session organizers who each p lanned a nd d eveloped a half-day session, invited the speakers, and presided over the session. The featured topics and the presiding organizers for each session were Ionic Liquid T utorials (J. F. Brennecke, J. D. Holbrey, R. D. Rogers, K. R. Seddon, and T. Welton), Fuels and Applications (J. H. Davis, Jr.) Physical and Thermodynamic Properties (L. P. N. Rebelo), Catalysis and Synthesis (P. Wasserscheid), Spectroscopy (S. Pandey), Separations (W. Tumas), Novel Applications (R. A. Mantz and P. C. Trulove), Catalytic Polymers and Gels (H. Ohno), Electrochemistry (D. R. McFarlane), Inorganic and Materials (M. Deetlefs and J. Holbrey), and General Contributions (R. D. Rogers and K. R. Seddon). Page restrictions prevented the publication of all the presentations (despite having the luxury of two volumes); however, we tried to select a representative subset of the papers.

Among the wonderful chemical contributions, there was one chastening note; a minute's silence was held at 8:46 on the morning of September 11th to remember those thousands of innocent people who had died exactly two years earlier very close to the Javits Conference Center in which we were holding the meeting. It was hard to escape the emotional impact of those events on that day.

The symposium was successful because of the invaluable support it received from industry, academia, government, and our professional society. Industrial support was received from Cytec Industries, Fluka, Merck KGaA/EMD Chemicals, SACHEM, Solvent Innovation, and Strem Chemicals. Academic contributions were received from The University of Alabama Center for Green Manufacturing and The Queen's University Ionic Liquid Laboratory (QUILL). The U.S. Environmental Protection Agency's Green Chemistry Program also supported the meeting. Of course, we are (as always) indebted to the ACS and its many programs for their help, encouragement, and support. We especially thank the ACS Division of Industrial and Engineering Chemistry, Inc., the I&EC Separations Science & Technology and Green Chemistry & Engineering Subdivisions, and the Green Chemistry Institute.

Another measure of success was the impressive strength of the student c ontributions, i n t he t utorial, o ral, and poster sessions; on this basis, our future is in safe hands. And, as is definitely true in a burgeoning field, the future is always more exciting than the past!!

Robin D. Rogers
Center for Green Manufacturing
Box 870336
The University of Alabama
Tuscaloosa, AL 35487
Telephone: +1 205–348–4323
Fax: +1 205–348–0823
Email: RDRogers@bama.ua.edu
URL: http://bama.ua.edu/~rdrogers/

Kenneth R. Seddon
QUILL Research Centre
The Queen's University of Belfast
Stranmillis Road
Belfast, Northern Ireland BT9 5AG
United Kingdom
Telephone: +44 28 90975420
Fax: +44 28 90665297
Email: k.seddon@qub.ac.uk
URL: http://quill.qub.ac.uk/

Contents (SS 902)

Applications

Structure and Spectroscopy

Chapter 1

Investigation of the Structure of 1-Butyl-3-methylimidazolium Tetrafluoroborate and Its Interaction with Water by Homo- and Heteronuclear Overhauser Enhancement

Andrea Mele

Dipartimento di Chimica, Materiali e Ingegneria Chimica "Giulio Natta" Politecnico di Milano, Via L. Mancinelli, 7 I–20131 Milano, Italy

The application of some NMR techniques based on homonuclear $^1H\{^1H\}$- and heteronuclear $^1H\{^{19}F\}$-NOE are illustrated for the room temperature ionic liquid 1-butyl-3-methylimidazolium tetrafluoroborate $[bmim]^+$ $[BF_4]^-$ in the presence of known amount of water. The experimental results on this model system provide information on cation-cation, cation-anion, water-cation and water-anion interactions. The results are discussed in terms of non-covalent C–H···O and C–H···F and O–H···F interactions. The results are compared to experimental and theoretical data taken from the most recent literature on similar systems.

The use of room temperature ionic liquids (RTILs) as non-volatile reaction media for synthesis and catalysis is constantly gaining popularity in both academia and industry (1-4). Nevertheless, their unique physical and physicochemical properties, as well as the mechanism of interaction with molecular solutes, are far from being fully understood. This, in turn, can severely hamper one of the most fascinating features of RTILs, that is the possibility of "designing" specifically targeted new solvents due to the large number of combinations of anions and cations potentially able to afford new RTILs (5, 6). The rational design of new RTILs, and the choice of optimum RTIL for a given synthetic or catalytic process, require the knowledge of the structure of the liquid at atomic level in terms of type and intensity of intermolecular interactions within the liquid phase, the interaction sites and the way the intermolecular interactions affect or determine the behavior of the RTIL in the presence of a given solute. In a schematic way, a detailed study of the structure of RTILs should provide data concerning cation-cation, cation-anion, solute-cation and solute-anion interaction.

Special interest is aroused by water as solute. This is basically due to the fact that many RTILs, either hydrophilic or hydrophobic, are known to absorb relevant amount of water from the environment. It is proven that important physicochemical properties of RTILs are dependent on the amount of the absorbed water. A detailed description of the influence of dissolved water on viscosity, conductivity, polarity, rate and selectivity of chemical reactions carried out in ionic liquids is beyond the purpose of this chapter. For the sake of clarity suffice it to mention the variation of RTILs' polarity (7) and the change of CO_2 solubility in RTILs (8) in the presence of water. The latter parameter is of interest for the application of carbon dioxide for the separation of RTILs from organic mixtures (9, 10).

The purpose of this chapter is to illustrate the application of some NMR methods based on the detection of homonuclear and heteronuclear Overhauser enhancement (or effect, NOE) to the structural problems mentioned above. To this end, some applications of $^1H\{^1H\}$- and $^1H\{^{19}F\}$-NOE experiments in one and two dimensions for the investigation of 1-butyl-3-methylimidazolium tetrafluoroborate [bmim]$^+$ [BF$_4$]$^-$ (compound 1, Figure 1) and its interaction water (11) are exposed in detail as a case study. The discussion is largely focused on the aspects concerning the repertoire of intermolecular interactions consistent with the experimentally observed NOEs, such as "weak" hydrogen bonds (12), as possible source of local structure of the liquid and their role in the mechanism of interaction with water.

The experimental results are discussed and compared to other experimental and theoretical data of the most recent literature on the same systems or on structurally related RTILs (Figure 1).

The Structure of RTILs by NMR spectroscopy: Applications of Nuclear Overhauser Enhancement

The physical principles of the nuclear Overhauser effect, the applications of NMR experiments based on NOE and the details of the pulse sequences used for NOE experiments are extensively reported in fundamental textbooks (*13-15*) and will not be repeated here. In the first part of this section some examples of application of two-dimensional homonuclear NOE correlation spectroscopy in the laboratory frame (NOESY) and in the rotating frame (ROESY) are proposed. The latter is often used in order to avoid the region of vanishing NOE corresponding to the condition $\omega \tau_c \approx 1$, where ω is the Larmor frequency and τ_c is the rotational correlation time.

The first issue to be addressed when investigating the structure of ionic liquids thorough the NOE approach concerns the detection and the interpretation of inter-cation NOEs. This point was elegantly discussed by Osteryoung and coworkers (*16*) by using two-dimensional ROESY experiments on compound **7** in the presence of AlCl$_3$. The ROESY spectrum of a mixture 1.0:0.5 of [emim]$^+$ [Cl]$^-$ and AlCl$_3$ showed cross peaks due to NOE transfer from each proton to all the others. This is clearly a consequence of the superimposition of *intramolecular* and *intermolecular* contributions. The same experiment was repeated by using a mixture containing 95% perdeuterated **7** and 5% non-deuterated **7**. The effect is that, on average, each non-deuterated molecule is completely surrounded by perdeuterated molecules. This makes the distance of two non-dueterated molecules beyond the threshold of 4 Å commonly accepted for vanishing dipolar interactions. The ROESY spectrum of this mixture indeed displayed the intramolecular contributions only, thus allowing working out the intermolecular interactions by comparison of the cross peaks pattern with the previous experiment. The authors thus demonstrated that cross peaks connecting, for example, the pairs H2-H4, H2-H5, H4-H6 etc. are due to intermolecular NOE transfer. The presence of such effects provided evidence of a structure of the liquid. The authors invoked cation-anion hydrogen bonding as a rationale.

The NOE experiments for the assessment of the cation-anion interactions can be conveniently applied in RTILs with polyatomic anions containing suitable nuclei (e.g. ^{1}H, ^{19}F). In the case of RTILs containing monoatomic anions, halides typically (excluding F$^-$), the approach based on inter-ion NOE cannot be applied, and other NMR parameters, such as the chemical shifts of the interacting nuclei, were found to be sensitive probes for cation-anion interactions, as reported for compounds **7**, **8** and **9** (*17*).

Dupont *et al.* studied the structure of [bmim]$^+$ [B(C$_6$H$_5$)$_4$]$^-$ (compound **12**) in [D$_6$]-DMSO and CDCl$_3$ by NOESY (*18*). The experiment carried out in the latter solvent showed intermolecular contacts involving H2, H4, H5, H6 and H10 of the cation (see Figure 1 for atom numbering) and the aromatic hydrogen atoms of the phenyl rings of the anion. The results indicate the existence of contact ion pairs in chloroform solution with participation of interionic C–H···π interactions (Figure 2), further confirmed by X-ray crystallography.

RTILs containing fluorinated polyatomic anions such as BF_4^-, PF_6^-, $CF_3SO_3^-$, $(CF_3SO_2)_2N^-$ etc., are potential candidates for the detection of interionic heteronuclear $^1H\{^{19}F\}$-NOE. In its easiest form, such experiment can be carried out as the heteronuclear equivalent of the routine $^1H\{^1H\}$-NOE difference spectroscopy commonly used for structure assessment of organic molecules (19). In practice, the NMR instrument transmitter is set on the proton frequency; a second rf-channel is set on the ^{19}F frequency. Pre-irradiation of several seconds at low power on-resonance on the ^{19}F frequency of the anion is used to saturate the transitions and allow for $^1H\{^{19}F\}$-NOE development. A second experiment is carried out with the same settings but with off-resonance irradiation. The two experiments (on- and off-resonance) are interleaved for signal averaging during overnight accumulation. The reference spectrum (off-resonance) is then subtracted from the on-resonance. The difference spectrum is expected to show only the proton signals corresponding to the hydrogen atoms of the cation in spatial proximity to the saturated fluorine. Alternatively, a two dimensional experiment, $^1H\{^{19}F\}$ heteronuclear Overhauser effect correlation spectroscopy, ($^1H\{^{19}F\}$-HOESY) can be used (20). This experiment can be regarded as the heteronuclear analogue of NOESY.

The integrated approach based on the homo- and heteronuclear NOE is illustrated in the next section for the model system [bmim]$^+$ [BF$_4$]$^-$ / water. The homonuclear experiments (ROESY) are applied to the study of cation-cation and cation-water interactions, whilst the complementary information on cation-anion and anion-water interactions is obtained via the heteronuclear experiments (NOE difference and HOESY).

The Interaction of [Bmim]$^+$ [BF$_4$]$^-$ with Water

Recent structural studies approached the state of water in archetypal RTILs by using vibrational spectroscopy. Attenuated total reflectance (ATR) and transmission IR spectroscopy study of the v_3 and v_1 mode of vibration of H_2O dissolved in compounds **1-6** pointed out that water is present in such liquids as "free" water (*i.e.* not as bulk water) hydrogen bonded to the anion in a symmetric complex of type A^-···H–O–H···A^- (21). Tran *et al.* used near infrared (NIR) spectroscopy for determining the concentration and the state of absorbed

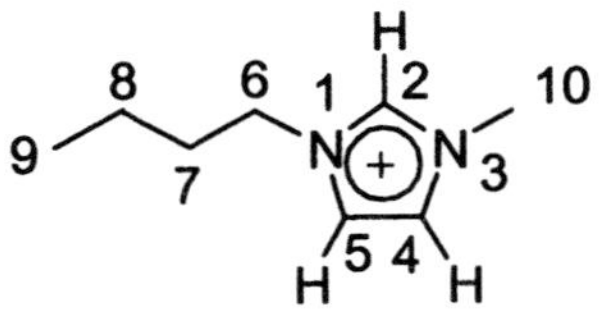

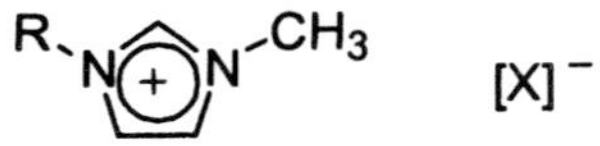

1	R=*n*-butyl	X=BF_4	**7**	R=Et	X=Cl
2	R=*n*-butyl	X=PF_6	**8**	R=Et	X=Br
3	R=*n*-butyl	X=SbF_6	**9**	R=Et	X=I
4	R=*n*-butyl	X=ClO_4	**10**	R=Me	X=Cl
5	R=*n*-butyl	X=CF_3SO_3	**11**	R=H	X=CF_3SO_3
6	R=*n*-butyl	X=$(CF_3SO_2)_2N$	**12**	R=*n*-butyl	X=$B(C_6H_5)_4$
			13	R=Me	X=PF_6

Figure 1.Top: Molecular structure and atom numbering of [bmim]$^+$. Bottom: scheme of RTILs cited throughout the text.

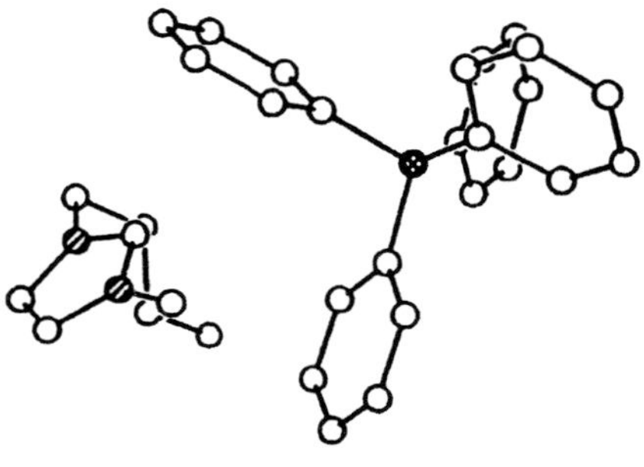

Figure 2. Scheme of contact ion pair of 12. (Reproduced with permission from reference 18. Copyright 2000 Wiley–VCH Verlag GmbH & Co.)

water in compounds **1, 2** and **6**. Position, shape and relative intensity of O–H overtones and combination bands of the added water were strongly dependent on the nature of anion of the RTIL, thus suggesting the presence of hydrogen bonds between water and the counter ion of the 1-butyl-3-methylimidazolium cation (*22*).

As a matter of fact, IR spectroscopy is still a classic method for the study of hydrogen bond (*23*). Nevertheless, the NMR approach based on NOEs represents a powerful complementary tool to IR spectroscopy: the mechanism of NOE is based on the cross-relaxation of dipolar-coupled nuclei and it is, therefore, a pair interaction. When applied to systems where non-conventional and weak hydrogen bonds (e.g. C–H···O–H; C–H···F–B or O–H···F–B) may take place, intermolecular NOE provides information on both hydrogen bond donor and acceptor. Also, the detection of intermolecular NOEs between water and cation and anion of model RTILs allows for the localization of the site of interaction. Examples are provided in the following sections.

Cation–Cation NOEs in [Bmim]$^+$ [BF$_4$]$^-$ in the presence of water

The samples for the ROESY experiments were prepared by adding known amounts of HPLC-grade water to pure **1**. The analytical data are reported in Table I. For each sample a ROESY spectrum was recorded and the intensity of cross peaks due to intermolecular interactions was evaluated by volume integration. The cross peak due to intramolecular H2-H10 interaction was set = 1 (arbitrary units) and used as reference. The relative intensity (%) is calculated as $[(V-V_0)/V_0]*100$, where V_0 is the volume of a given cross peak in pure **1**, whilst V is the volume of the same cross peak in the sample with added water. Negative values mean decreased NOE intensity with increasing water content, the vice versa for positive values.
The results, summarized in Figure 3, can be interpreted as follows:

- The relative intensity of cross peaks relating to the pairs H2-H4, H2-H5 and H5-H10 shows a systematic decrease as the water content is increased.
- The opposite tendency is observed for the intermolecular NOEs involving the methyl group in position 10 and the aliphatic protons of the butyl chain.

As introduced in the previous section, inter-cation NOEs are diagnostic of a local structure of the liquid, generally mediated by anion-cation hydrogen bonds. Considering the fact that NOE intensity is related to internuclear distance, the data of Figure 3 indicate that the distance between H2 of a given

8

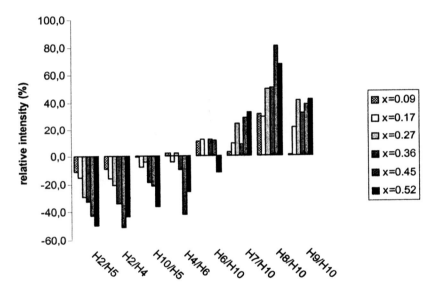

Figure 3. Histogram of the relative intensities of ROESY cross-peaks as function of water content in compound 1.The proton pairs giving rise to intermolecular NOE interaction are reported on the x-axis. (Adapted with permission from reference 11. Copyright 2003 Wiley–VCH Verlag GmbH & Co.)

[bmim]$^+$ and H5 of a neighboring cation is constantly growing by adding water to the system. The same holds for intermolecular distances associated to the other pairs H2-H4 and H5-H10. The dissolved water thus modifies the structure of the pure 1. A possible rationale is that water may compete with the aromatic protons as hydrogen bond donor toward the [BF$_4$]$^-$ anion. At the present moment there are experimental evidences that methylimidazolium RTILs with fluorinated anion posses a complex structure with participation of C–H$\cdots$F hydrogen bonds. These works are based on ^{13}C dipole-dipole relaxation rates (24) and diffusion coefficient measurement (25) on 8, ^{13}C relaxation data on 2 (26) and solvent effects on chemical shifts (27). A certain degree of stacking of the aromatic rings of the ionic liquid is thus a consequence of the hydrogen bond network between [bmim]$^+$ and [BF$_4$]$^-$.

Table I. Analytical data^a for NMR samples of [bmim]$^+$ [BF$_4$]$^-$.

	1°	1b	1c	1d	1e	1f	1g
1^b	3.76	3.84	3.86	3.79	3.84	3.86	3.82
H$_2$O^b	0	0.39	0.78	1.39	2.17	3.11	4.17
n(H$_2$O)/n(1)^c	0	0.10	0.20	0.37	0.56	0.81	1.09
x(H$_2$O)^d	0	0.09	0.17	0.27	0.36	0.45	0.52
m(H$_2$O)	0	0.45	0.89	1.62	2.50	3.57	4.83

a (Adapted with permission from reference 11. Copyright 2003 Wiley–VCH Verlag GmbH & Co.)

b In mmol.

c Mole ratio water/ionic liquid.

d Water molar fraction.

e Water molality.

This kind of organization accounts for the observed intermolecular NOEs of type H2-H4, H2-H5 and H5-H10. By adding water, the local structure is modified to a different degree and type of cation-cation organization with consequent increase of the distance between the protons mentioned above. Concerning the nature of such organization, "stacking" of the aromatic rings, or some other kind of piling, could be reasonable hypotheses. Remarkable examples of aromatic ring cation-cation stacking are reported for compounds 7 and 8 by NMR (17) and for 11 by X-ray crystallography (28).

The trend of cross-peak intensity related to H6 intermolecular contacts is less clear, possibly due to the conformational flexibility of the butyl chain.

The aliphatic hydrogen H7, H8atoms show a complementary behavior. The general trend observed is consistent with the minimization of water / aliphatic chains interactions and maximization of the hydrophobic interactions between methyl 10 and the butyl chain. The role of hydrophobic interactions as stabilization factor in [bmim]$^+$ chloride was recently proposed by Hamaguchi and coworkers (29).

Water–Cation NOEs in [bmim]$^+$ [BF$_4$]$^-$

The relative intensity of cross peaks related to water – 1 NOEs is reported in Table II. The examination of the data point out that:

- At low water content (samples 1b and 1c, corresponding to cation to water ratio of 10:1and 5:1, respectively) intermolecular NOEs are detectable on the imidazolium ring protons only, H2 in particular. The water molecules are close in space to the aromatic ring protons H2, H5 and H4.
- With increasing water content all the other cross peak intensities rise, especially those related to H6 and H10.
- At higher water content the NOEs intensities do not show any selectivity (a part from the slightly lower values for cross peaks involving H7 and H8), thus suggesting a sort of solvation of the cation rather than a specific interaction with selected protons.

Table II. Volume integration^a of ROESY cross peaks between water (w) and imidazolium protons of [bmim]$^+$ [BF$_4$]$^-$

	1b	1c	1d	1e	1f	1g
w-H2	2.8	6.6	10.7	14.9	19.7	27.4
w-H5	2.6	3.5	7.9	9.0	17.5	18.4
w-H4	1.8	5.5	8.3	10.7	16.8	21.5
w-H6	0	4.1	7.5	12.9	17.8	25.6
w-H10	0	6.0	9.8	19.3	31.6	31.5
w-H7	0	1.5	5.1	6.8	12.0	17.5
w-H8	0	1.5	4.0	6.0	10.7	16.7
w-H9	0	2.3	5.6	6.9	14.4	24.3

a The volumes are in arbitrary units. The numerical values are scaled to the volume of H2-H10 cross peak arbitrarily set = 100 (Adapted with permission from reference 11. Copyright 2003 Wiley–VCH Verlag GmbH & Co.).

The role of water-cation interactions in [bmim]$^+$ based ionic liquids is controversial. Strong evidence that water is mainly interacting with the anion is provided by IR spectroscopy (*21, 22*). Molecular dynamics simulations of water dissolved in ionic liquids **10** and **13** pointed out the existence of strong hydrogen bonds of individual water molecules with the anion (*30, 31*). The next section of this chapter will add NOE evidence consistent with water-anion hydrogen bonds. Nonetheless, the observed NOEs between water and cation protons clearly indicate that water molecules are in close contact with the cation. This could be regarded as a simple consequence of the H_2O-$[BF_4]^-$ attractive interaction that forces water to surround the cation. However, a participation of H_2O-$[bmim]^+$ interactions to the stabilization of the system cannot be ruled out *a priori* and the possibility that $[bmim]^+$ hydrogen atoms could act as hydrogen bond donors toward water might be taken into account as a working hypothesis. Indeed, the experimental data reported above indicate that water molecules are in the spatial proximity of H2 methyl 10 and methylene 6 (Figure 4).

The intensity of NOEs related to the interaction of water with the other protons of the butyl chain (H7, H8 and H9) is low compared to that of H2, as displayed in Figure 5. The curves of Figure 5 again confirm the localization of the water interaction close to H2. Moreover, the data of Table II point out significant interaction of water with H4 and H5, although of minor intensity with respect to H2. These findings suggest that the aromatic ring protons (mainly H2) may either compete with the anion for binding water or be involved in multicentric interactions.

Cation–Anion and water–anion NOEs in $[bmim]^+$ $[BF_4]^-$ in the presence of water

Direct evidence of cation-anion hydrogen bonds can be obtained from heteronuclear NOE experiments. The contour plot of a two dimensional $^1H\{^{19}F\}$-HOESY experiment (mixing time 600 ms) carried out on sample **1g** is showed in Figure 6. The F1 projection clearly shows the presence of the isotopic species $[^{10}BF_4]^-$ (ca. 20% abundance) and $[^{11}BF_4]^-$. The cross peaks are due to heteronuclear NOE between the fluorine of the anion and the proton of the cation or water. The spectrum indicates that:

- Close contacts are present between the ring protons (H2, H4 and H5) and the fluorine atoms of the anion. The most intense interaction involves H2. Having set the cross peak volume of the H2/F NOE = 100 (a.u.), the cross peaks related to H4 and H5 integrate 81 and 77 in the same intensity scale, respectively.

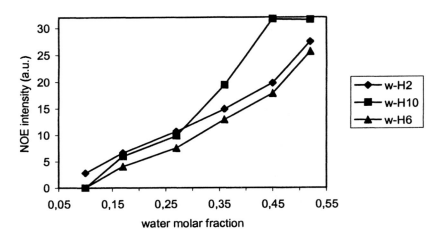

Figure 4. Relative intensity of ROESY cross peaks showing the interaction of water with H2, H6 and H10. (Adapted with permission from reference 11. Copyright 2003 Wiley–VCH Verlag GmbH & Co.).

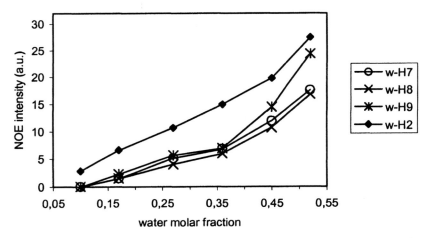

Figure 5. Relative intensity of ROESY cross peaks showing the interaction of water with aliphatic hydrogen atoms H7, H8 and H9. The curve related to water-H2 interaction is also reported for the sake of comparison. (Adapted with permission from reference 11. Copyright 2003 Wiley–VCH Verlag GmbH & Co.).

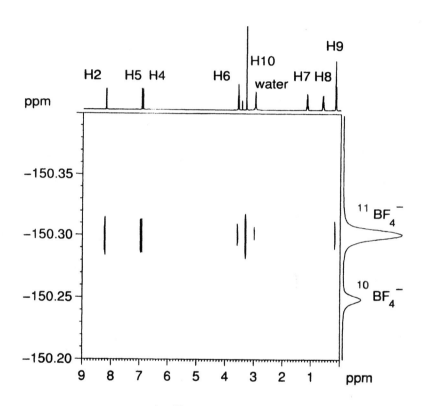

Figure 6. Contour plot of $^1H\{^{19}F\}$-HOESY carried out on compound 1g.

- The fluorine nuclei are close in space also to H6 and H10, namely the aliphatic hydrogen atoms located in the α-position with respect to imidazolium N atoms. However, the quantitation of the cross peak volume is affected by errors due to the presence of T1 noise.
- Small NOE is also observed on H9, namely the end group of the butyl chain. Again the quantitation is hampered by the presence of T1 noise.
- A significant cross peak water / $[BF_4]^-$ of volume = 80 (see above for units) is also detected.

The data are consistent with close contacts between the aromatic imidazolium protons and the fluorine atoms of the anion. This, in turn, is consistent with the presence of C–H···F–BF_3^- hydrogen bonds. The most intense interaction involves H2 as hydrogen bond donor. This finding is coherent with the observation, based on ^{13}C relaxation data, that the C(2)–H bond in [bmim]$^+$ [PF$_6$]$^-$ **2** undergoes severe restriction of molecular motion due to the formation of ion pairs or even larger aggregates (*26*). The geometry of ion pairs [bmim]$^+$–[PF$_6$]$^-$ was calculated *ab initio* at different levels of approximation, and showed formation of ion-pairs in the gas phase. The most important non-bonded interactions were C(2)–H···F–PF_5^- hydrogen bonds. Significant short contacts C–H···F were also reported for the methyl 10 and some of the methylene hydrogen atoms of the butyl chain, thus suggesting that also the aliphatic groups might play a role in the hydrogen bond donation (*32*). The same kind of participation of the aliphatic groups in the α-position with respect to imidazolium N atoms can be invoked to rationalize the NOEs of H10 and H6 with the fluorine nuclei of the anion above described for compound **1**.

Cation-anion aggregation is still present in dilute solution of (D_6)-DMSO (45 mg / 750 μL corresponding to ca. 0.3 M), as proven by ^{1}H{^{19}F} NOE difference experiments. The results, displayed in Figure 7, are consistent with close contact of type C–H···F–BF_3^- involving the aromatic protons of the cation. These results further support what already reported for more concentrated solutions (*11*) and provide evidence of ion-pair formation even in DMSO solution in the explored concentration range.

As mentioned before, the results ^{1}H{^{19}F}-HOESY experiment point out a significant close contact of water and F atoms of $[BF_4]^-$ (Figure 6). We interpret this finding as evidence of water–anion hydrogen bond. Water molecules thus seem to compete with the aromatic C–H bonds of [bmim]$^+$ cation as hydrogen bond donor toward the fluorinated anion.

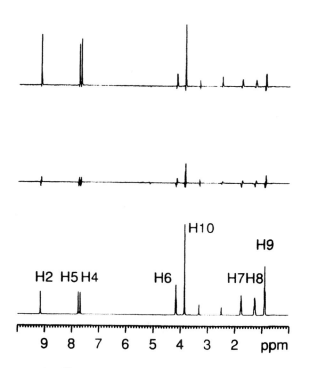

Figure 7. Top: of $^1H\{^{19}F\}$ NOE difference spectrum of 0.3 M solution of 1 in [D₆]-DMSO. Bottom: reference spectrum (off-resonance). Middle: example of subtraction of two different off-resonance spectra.

References

1. Welton, T. *Chem. Rev.* **1999,** *99,* 2071–2083.
2. Dupont, J.; de Souza, R. F.; Suarez, P. A. Z. *Chem. Rev.* **2002,** *102,* 3667–3692.
3. *Ionic Liquids in Synthesis;* Wasserscheid, P.; Welton, T., Eds.; VCH-Wiley: Weinheim, 2002.
4. *Ionic Liquids; Industrial Applications for Green Chemistry;* Rogers, R. D.; Seddon, K. R., Eds.; ACS Symposium Serie 818; American Chemical Society: Washington DC, 2002.
5. Freemantle, M. *Chem. Eng. News* **1998,** *76,* 32–37.
6. Earle, M. J.; Seddon, K. R. *Pure Appl. Chem.* **2000,** *72,* 1391-1398.
7. Aki, S. N. V. K.; Brenneke, J. F.; Samanta, A. *Chem. Commun.* **2001,** 413–414.
8. Blanchard, L. A.;Gu, Z.; Brenneke, J. F. *J. Phys. Chem. B* **2001,** 105, 2437–2444.
9. Scurto, A. M.; Aki, S. N. V. K.; Brenneke, J. F. *J. Am. Chem. Soc.* **2002,** *124,* 10276–10277.
10. Scurto, A. M.; Aki, S. N. V. K.; Brenneke, J. F. *Chem. Commun.* **2003,** 572–573.
11. Mele, A.; Tran, C. D.; De Paoli Lacerda, S. H. *Angew. Chem. Int. Ed. Engl.* **2003,** *42,* 4364–4366.
12. Desiraju, G. R.; Steiner, T. *The Weak Hydrogen Bond in Structural Chemistry and Biology;* IUCr Monographs on Crystallography 9; Oxford University Press: Oxford, UK, 1999.
13. Neuhaus, D.; Williamson, M. *The Nuclear Overhauser Effect in Structural and Conformational Analysis,* VCH: New York, 1989.
14. Ernst, R. R; Bodenhausen, G.; Wokaun, A. *Principles of Nuclear Magnetic Resonance in One and Two Dimensions;* International Series of Monographs on Chemistry 14; Oxford University Press: Oxford, UK, 1987.
15. Wüthrich, K. *NMR of Proteins and Nucleic Acids,* John Wiley & Sons: New York, 1989.
16. Mantz, R. A.; Trulove, P. C.; Carlin, R. T.; Osteryoung, R. A. *Inorg. Chem.* **1995,** *34,* 3846–3847.
17. Avent, G. A.; Chaloner, P. A.; Day, M. P.; Seddon, K. R.; Welton, T. *J. Chem. Soc. Dalton Trans.* **1994,** 3405–3412.
18. Dupont, J.; Suarez, P. A. Z.; De Souza, R. F.; Burrow, R. A.; Kintzinger, J.-P. *Chem. Eur. J.* **2000,** *13,* 2377–2381.

19. Derome, A. E. *Modern NMR Techniques for Chemistry Research;* Baldwin, J. E., Ed.; Organic Chemistry Series 6; Pergamon Press: Oxford, 1987.

20. Gerig, J. T. *Magn. Reson. Chem.* **1999,** *37,* 647–652 and references therein.

21. Cammarata, L.; Kazarian, S. G.; Salter, P. A.; Welton, T. *Phys. Chem. Chem. Phys.* **2001,** *3,* 5192–5200.

22. Tran, C. D.; De Paoli Lacerda, S. H.; Oliveira D. *Appl. Spectrosc.* **2003,** *57,* 152–157.

23. Jeffrey, J. A. *An Introduction to Hydrogen Bonding,* Oxford University Press: Oxford, UK, 1997; pp 220–225.

24. Huang, J.-F.; Chen, P.-Y.; Sun, I-W.; Wang, S. P. *Inorg. Chim. Acta* **2001,** *320,* 7–11.

25. Noda, A.; Hayamizu, K.; Watanabe, M. *J. Phys. Chem. B* **2001,** *105,* 4603–4610.

26. Antony, J. H.; Mertens, D.; Dölle, A.; Wasserscheid, P.; Carper, W. R. *Chem. Phys. Chem.* **2003,** *4,* 588–594.

27. Headley, A. D.; Jackson, M. N. *J. Phys. Org. Chem.* **2002,** *15,* 52–55.

28. Wilkes, J. S.; Zaworotko, M. J.; *Supramol. Chem.* **1993,** *1,* 191–193.

29. Saha, S.; Hayashi, S.; Kobayashi, A.; Hamaguchi, H. *Chem. Lett.* **2003,** *8,* 740–741.

30. Hanke, C. G.; Atamas, N. A.; Lynden-Bell, R. M. *Green Chem.* **2002,** *4,* 107–111.

31. Hanke, C. G.; Lynden-Bell, R. M. *J. Phys. Chem. B* **2003,** *107,* 10873–10878.

32. Meng, Z.; Dölle, A.; Carper, W. R. *J. Mol. Struc. (THEOCHEM)* **2002,** *585,* 119–128.

Chapter 2

Apples to Apples: A Comparison of Lanthanide β-Diketonate Complexes in Molecular Solvents and an Ionic Liquid

Mark P. Jensen, James V. Beitz, Jörg Neuefeind, S. Skanthakumar, and L. Soderholm

Chemistry Division, Argonne National Laboratory, Argonne, IL 60439

The structure, stoichiometry, and chemical equilibria governing the complexes formed by lanthanide ions (Ln^{3+}) and the extractant 2-thenoyltrifluoroacetone (Htta) were studied in a biphasic system composed of aqueous 1 M $NaClO_4$ and the hydrophobic ionic liquid, 1-butyl-3-methylimidazolium bis(trifluoromethylsulfonyl)imide. Two different Ln-tta complexes were observed in the ionic liquid phase depending on the solution conditions. A neutral $Ln(tta)_3$ complex predominates at low tta^- concentration, while the anionic complex $Ln(tta)_4^-$ is extracted into the ionic liquid phase by an anion exchange mechanism at higher tta^- concentrations. Both complexes have been observed in molecular solvents previously, allowing a direct and quantitative comparison of the ionic liquid based system to conventional liquid-liquid extraction systems.

Liquid-liquid extraction is an important process for separating and purifying metal ions that is easily scalable from benchtop experiments up to plant-scale industrial hydrometallurgy. The chemical and engineering aspects of liquid-liquid extraction with conventional molecular organic solvents are well understood, but the need to build more environmentally friendly processes has lead to the consideration of several alternate solvent systems for metal ion extraction (1-3). Among the replacements for conventional organic solvents being investigated, water-immiscible room-temperature ionic liquids (RTILs) appear to be the most compatible with current approaches to solvent extraction based metal ion separations. However, the relevant solvent properties of only a few RTILs are known, and the data required for a systematic understanding of the extraction properties of metal complexes in biphasic water-RTIL systems are non-existent. The development of RTIL-based metal ion separations processes will be slow without this information. Given the immense existing knowledge base of liquid-liquid extraction in conventional solvent systems, quantitative comparisons between extraction in ionic liquid-based systems to the extraction observed in molecular solvent systems could be very useful.

Three key capacities are required to develop predictive capabilities for metal ion extraction by RTILs. First, the physico-chemical properties of the RTILs themselves must be systematically understood. This includes measurements of properties such as polarity, viscosity, water-miscibility, and various solvation parameters. Some of the earliest work suggesting that certain RTILs might be good solvents for liquid-liquid extraction began with this approach (1,4). Such parameters also are central to the rational application of RTILs in many synthetic or catalytic systems. Consequently, work is underway to measure and systematize RTIL properties. Second, the chemical equilibria involved in the partitioning of solutes (especially complexes of metal ions) in biphasic RTIL systems must be understood. Some early research in this area has identified unique equilibria in RTILs that are not possible in extraction systems using purely molecular organic solvents (5). Third, the coordination chemistry of the metal ion complexes present in the RTIL-phase must be characterized. Work in this regard has been done in chloroaluminate melts, but is only beginning in biphasic RTIL systems.

The comparison of the extraction of metal complexes into RTILs vis-à-vis conventional solvents is difficult because many more possible chemical equilibria may be operative in RTIL systems. More importantly though, the ionic nature of the RTIL phase can encourage the formation of ionic metal complexes which have no known analog in molecular solvent systems (6,7). The very efficient extraction of Sr-crown ether complexes (5,8-10) as compared to that observed in the molecular solvent, 1-octanol, was shown to arise from a change in the extraction mechanism that enabled the formation of a new complex in the RTIL phase (5,7). In order to quantitatively compare the extraction of metal complexes one must compare the same equilibria, but to this point all of the extraction equilibria of metal complexes that have been fully investigated involve equilibria or complexes that are unique to the RTIL systems.

To shed light on each of the three key capacities, we have studied the extraction of trivalent lanthanide cations (Ln^{3+}) from a 1.0 M (H,Na)ClO$_4$ aqueous phase into an equal volume of the hydrophobic ionic liquid 1-butyl-3-methylimidazolium bis(trifluoromethylsulfonyl)imide ($C_4mim^+Tf_2N^-$) in the presence of the chelating β-diketone extractant 2-thenoyltrifluoroacetone (Htta, Figure 1) at 22 ± 2 °C. The behavior of Ln^{3+}-tta$^-$ complexes in molecular

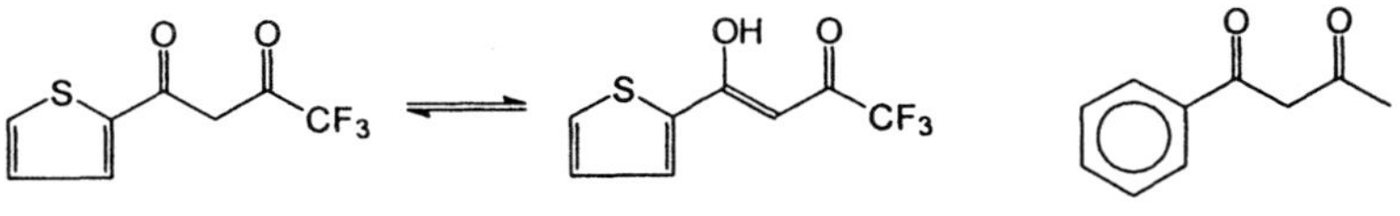

2-Thenoyltrifluoroacetone, Htta   1-Benzoylacetone, Hbza

Figure 1. Structures of β-diketones and the keto-enol tautomers of Htta

solvents is both well established and varied. Neutral, cationic, and anionic Ln^{3+}-tta$^-$ complexes have all been reported in conventional solvents (11-15). Therefore this chemical system presents an opportunity to develop a comparative understanding of chemical equilibria in biphasic water-RTIL systems. Our experiments highlight the unique nature of ionic liquids as solvents for metal ion extraction. With respect to the partitioning of the extractant itself, the solvent properties of $C_4mim^+Tf_2N^-$ could be considered to resemble relatively non-polar molecular solvents like benzene or chloroform or more polar solvents like 1-hexanol. In the presence of Eu^{3+} or Nd^{3+}, the metal distribution ratios imply that two different Ln^{3+}-tta$^-$ complexes are present in the RTIL phase depending on the concentration of tta$^-$. The coordination environments of the two complexes were studied and both a neutral complex, $Ln(tta)_3(H_2O)_m$, and an anionic complex, $Ln(tta)_4^-$, are present in the RTIL phase even although neither species is observable in the aqueous phase. The change in the extraction mechanism from partitioning of the neutral $Ln(tta)_3(H_2O)_m$ complex to anion exchange of $Ln(tta)_4^-$ for Tf_2N^- is driven by the Htta concentration and aqueous acidity. With direct identification of the extracted species and the equilibria involved in the partitioning reaction, the extraction of the Ln-tta complexes into $C_4mim^+Tf_2N^-$ can be quantitatively compared to that in molecular solvent systems.

Partitioning in the Absence of Lanthanide Cations

Recent work in liquid-liquid extraction of metal ions into ionic liquids by various ligands has demonstrated that unusual extraction mechanisms exist in some RTIL systems. Nevertheless, the measurement of metal ion partitioning in RTILs is made more difficult because identification and quantification of the equilibria involved in the partitioning of *metal complexes* between two liquid phases often requires an understanding of the partitioning equilibria of the other components. The liquid-liquid extraction system under consideration is comprised of H_2O, $NaClO_4$, and $HClO_4$, in addition to the obviously interesting components, Htta, Ln^{3+} (with three ClO_4^- counterions), and $C_4mim^+Tf_2N^-$. The partitioning of the "background" ionic species Na^+, H^+, and ClO_4^- from the aqueous phase into non-polar, molecular organic solvents as ion pairs is typically low and, in any event, only neutral ion pairs can be present in the organic phase at equilibrium. The presence of β-diketone extractants does little to change this situation. RTILs, in contrast, are ionic and somewhat polar solvents (*16*). Consequently, ionic liquids might dissolve significant concentrations of ionic species, either as ion pairs or as individual, "free" ions through exchanging C_4mim^+ for cations or Tf_2N^- for anions.

Combining analytical measurements of the H_2O, Na^+, H^+, and C_4mim^+ concentrations with the requirements of charge neutrality within each phase and mass balance in the system, the equilibrium concentration of each of the solutes was determined in the absence of lanthanide cations. Further experimental details are given by Jensen et al. (*17*). As summarized in Table I, only small amounts of any of these solutes are transferred from one phase to the other in the absence of lanthanide cations. The RTIL-phase had a small proton capacity (ca. 0.001 M) when the aqueous phase pcH approached 3, but it is not clear if this is an inherent property of this RTIL, or if it arises from a small amount (<0.1 %) of a weak Brønsted base impurity in the RTIL.

Table I. Equilibrium Composition of Equal Volumes of Aqueous 1 M NaClO$_4$ at pcH = 2 and 0.499 M Htta in C$_4$mim$^+$Tf$_2$N$^-$

Species	Aqueous Phase, moles/L	RTIL Phase, moles/L
Na^+	0.99	0.0038
H^+	0.010	< 0.0002
Htta	0.0107	0.488
tta$^-$	5.6×10^{-7}	Not Detected [a]
C_4mim^+	0.034	3.17
Tf_2N^-	0.034 [b]	3.17
ClO_4^-	1.00	0.0038 [c]
H_2O	53	1.05

[a]Limit of detection is 4×10^{-4} M when the aqueous tta$^-$ concentration is 0.003 M
[b]Estimated from the [C_4mim^+] and [Tf_2N^-] equivalence in a 1.0 M NaCl system (*17*)
[c]From mass and charge balance. [ClO_4^-]$_{max}$ = 0.0040 M, considering [H^+]$_{max}$ = 0.0002 M

In conventional liquid-liquid extraction systems and in the absence of significant concentrations of metal cations, the concentration of a β-diketone extractant like Htta in either phase is controlled by two equilibria:

$$\text{Htta} \rightleftharpoons \overline{\text{Htta}} \tag{1}$$

$$\text{Htta} \rightleftharpoons \text{H}^+ + \text{tta}^-. \tag{2}$$

(The bar over a species indicates that it is present in the organic or RTIL phase.) Eq. 1 is described by the distribution constant K_d, which is affected both by the identity of the organic solvent and, to a lesser extent, by the ionic strength of the aqueous phase. Eq. 2 is the acid dissociation of Htta in the aqueous phase, represented by the acid dissociation constant, K_a, which only depends on the composition of the aqueous phase. Previously, the pK_a of Htta in 1 M NaClO$_4$ was reported to be 6.28 (18), and this value was used in all calculations. In an RTIL-based system, however, the extractant also could exist as tta$^-$ anions in the RTIL-phase, which would require additional equilibria to describe the system.

To probe this possibility, the distribution constant of Htta, purified by vaccum sublimation, was measured as a function of the Htta concentration and pcH by UV spectrophotometry and as a function of the pcH (pcH = -log [H$^+$] in molarity) by the increase in the titratable protons in the aqueous phase after equilibration. Taken together, these experiments allow one to follow the partitioning of Htta and tta$^-$ specifically. The results, summarized in Table II, give an average log K_d = 1.66 $\pm$ 0.01. The equivalence of the K_d values measured spectrophotometrically at pcH 3.03 and 5.67 and the values measured in more acidic solutions by titration indicate that Eq. 1 and 2 account for at least 99.9% of the extractant molecules in both phases, and that appreciable amounts of anionic tta$^-$ do not exist in the RTIL phase.

Table II. K_d of Htta in the 1 M (Na,H)ClO$_4$/C$_4$mim$^+$Tf$_2$N$^-$ System

Initial [$\overline{Htta}$], M	Equilibrium pcH	K_d	Technique (No. Measurements)
0.010 – 0.501	5.67	45.4 $\pm$ 0.7	Spectroscopy (6)
0.499	-0.01[a]	25 $\pm$ 20	Acid Titration (1)
0.499	1.00	44.4 $\pm$ 0.8	Acid Titration (1)
0.499	2.00	44.4 $\pm$ 0.8	Acid Titration (1)
0.509	3.03	45.2 $\pm$ 0.9	Spectroscopy (1)

[a] [HClO$_4$] = 1.03 M at equilibrium

From the perspective of Htta, the RTIL behaves like a molecular solvent. The affinity of Htta for the RTIL phase, represented by the K_d, can be compared to its affinity for relatively non-polar, molecular aromatic solvents such as benzene (Log K_d = 1.73 for 1 M NaClO$_4$ (18)). It can also be considered to be intermediate between the affinity of Htta for alcohols (e.g., Log K_d = 1.46 for 1-octanol and 0.1 M NaClO$_4$ (19)) and for esters and ketones (Log K_d = 1.8 – 2.3 for 0.1 M NaClO$_4$ (19)). Since Htta possesses both polar and non-polar functionalities, it is not surprising that its affinity for $C_4mim^+Tf_2N^-$ mimics a variety of conventional molecular solvents, and that a single parameter is not sufficient to compare its behavior (20).

Identity and Partitioning of the Lanthanide-tta Complexes

With an understanding of the partitioning of the other species in solution, the partitioning of the Ln^{3+}-tta$^-$ complexes between the aqueous and RTIL phases can be studied quantitatively. Slope analysis is one simple method for probing the general extraction equilibrium in this extraction system,

$$Ln(H_2O)_h^{3+} + n\,\overline{Htta} \rightleftharpoons \overline{Ln(tta)_x(Htta)_{n-x}(H_2O)_m} + x\,H^+ + (h-m)\,H_2O, \quad (3)$$

which is described by the extraction constant, K_{ex}. The extraction constant can be related to the lanthanide distribution ratio ($D_{Ln} = [\overline{Ln}]/[Ln]$), by the equation

$$Log\,D_{Ln} = n\,Log\,[\overline{Htta}] + x\,pcH + Log\,K_{ex}. \quad (4)$$

The slope of the Log D_{Ln} vs Log $[\overline{Htta}]$ curve at any given point gives n, the average tta:Ln stoichiometry of the extracted complex, while the slope of the Log D_{Ln} vs. pcH curve, x, equals the number of tta$^-$ anions in the organic phase complex. When the pcH of a set of samples varies, the aqueous concentration of tta$^-$ can be substituted for the organic phase Htta concentration to put the data on the same scale. If the extractant concentration and the pcH are known, the extraction constant, K_{ex}, can be calculated from the distribution data as well.

In a molecular organic solvent, hydrated, neutral Ln^{3+}-tta$^-$ complexes are generally extracted as $Ln(tta)_3(H_2O)_m$ (11). The stoichiometric parameters from Eq. 3 that describe this are $n = x = 3$ and $m = 1, 2$ or 3. Other neutral complexes, such as $Ln(tta)_3(Htta)$ ($n = 4$, $x = 3$, $m = 0$), also are observed at high extractant concentrations in some solvents (*21*), and cationic $Ln(tta)_2^+$ or anionic $Ln(tta)_4^-$ can be observed in some organic solvents in the presence of low charge density counter ions such as ClO_4^- or $(C_4H_9)_4N^+$. In the absence of Htta, appreciable amounts of Ln^{3+} are not extracted into $C_4mim^+Tf_2N^-$ (Log D_{Eu} = -3.82). In contrast, Figure 2 shows that in the presence of Htta, two complexes with different Eu:tta stoichiometries can be extracted into $C_4mim^+Tf_2N^-$. At lower extractant concentrations or pcH values, least-squares slope analysis the data with a two complex model indicates that three tta$^-$ molecules (*n* = 3.1 ± 0.2) are involved in the extraction of each Eu^{3+} ion into the RTIL. At higher concentrations ([tta$^-$] > 1 x 10^{-6} M), however, the slope of the curve increases to 4.0 ± 0.4, indicating that four tta$^-$ molecules are bound to each Eu^{3+} ion extracted under those conditions.

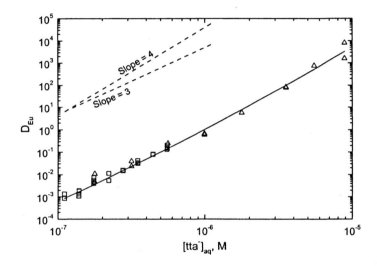

Figure 2. Eu^{3+} extraction into Htta/C$_4$mim$^+$Tf$_2$N$^-$ from 1.0 M (Na,H)ClO$_4$ (□) pcH = 2, (△) pcH = 3, (—) best fit to a 1:3 and a 1:4Ln:tta$^-$ complex.

To confirm these findings and fully identify the extraction mechanisms involved in the formation of the RTIL phase complexes, the coordination environments of the Ln^{3+}-tta$^-$ complexes present in the RTIL phase were studied by optical absorption and fluorescence spectroscopies, high energy X-ray scattering (HES), EXAFS, and by further extraction equilibrium measurements. The lanthanide Nd was used because it has stronger optical absorption bands, while Eu was used for its superior fluorescence properties and the easy availability of Eu radioisotopes. The low metal ion distribution ratios in the tta$^-$ concentration range that favors the formation of the 1:3 Ln:tta complex prohibited additional experiments with Htta at those concentrations. However, the Ln^{3+} extraction properties of the related β-diketone, 1-benzoylacetone or Hbza (Figure 1), allowed the extraction of sufficient Nd into the RTIL phase to study the optical absorption spectrum of the 1:3 Nd:bza complex in $C_4mim^+Tf_2N^-$. The spectrum of the 1:3 Nd:bza complex in $C_4mim^+Tf_2N^-$ matches the spectrum of $Nd(tta)_3(H_2O)_3$ in o-xylene (Figure 3). This is a strong indication that the coordination environment of Nd^{3+} in the two complexes is the same, and that the complex $Nd(bza)_3(H_2O)_3$ is extracted into $C_4mim^+Tf_2N^-$. Given the similarities of Hbza and Htta as ligands and the 1:3 Eu:tta ratio observed in $C_4mim^+Tf_2N^-$ at low Htta concentrations, the 1:3 complex present in the RTIL phase is most likely $Eu(tta)_3(H_2O)_3$ — the same neutral complex commonly observed in non-polar molecular organic solvents.

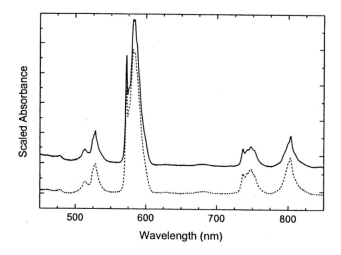

Figure 3. Absorption spectra of $Nd(tta)_3(H_2O)_3$ in o-xylene (solid line) and of the Nd-benzoylacetonato complex in 0.2 M $Hbza/C_4mim^+Tf_2N^-$ (dashed line)

In contrast to the difficulties making spectroscopic measurements on solutions of the 1:3 Ln:tta complex in $C_4mim^+Tf_2N^-$, solutions containing sufficient Eu or Nd concentrations for spectroscopic or scattering measurements were readily prepared under the conditions that favor the formation of the 1:4 Ln:tta complex in the extraction experiments. The optical absorption spectrum of the Nd-tta complex in the RTIL at high tta^- concentrations matches the spectrum of $Nd(tta)_4^-$ in conventional solvents containing a liquid anion exchanger (e.g., an organophilic tetraalkylammonium salt). Luminescence lifetime measurements of the Eu complexes in the water-equilibrated RTIL phase are also consistent with the presence of anionic $Ln(tta)_4^-$ complexes. The average number of water molecules in the Eu inner coordination sphere is proportional to the difference in the Eu^{3+} luminescence decay constants in H_2O and D_2O containing solutions (22). For Eu extracted into 0.5 M $Htta/C_4mim^+Tf_2N^-$, 0.1 ± 0.5 water molecules (calculated from $k_{H2O} = 2.38$ $msec^{-1}$ and $k_{D2O} = 2.29$ $msec^{-1}$ (17)) were detected coordinated to the Eu. This implies that the inner coordination sphere of the completely dehydrated Ln cations is filled by the eight oxygen atoms of four bidentate tta⁻ anions.

Measurements using synchrotron X-rays provide even more certain identification of the complex present in the RTIL phase at high tta⁻ concentrations. The L_3-edge EXAFS of the Eu complex extracted into 0.2 M $Htta/C_4mim^+Tf_2N^-$ indicate the presence of 8.7 ± 1.0 oxygen atoms in the Eu inner coordination sphere at an average Eu-O distance of 2.39 ± 0.01 Å. HES measurements, which generally give more accurate coordination numbers, but less accurate bond distances than EXAFS, were made on an RTIL solution with higher Eu loading (0.046 M Eu/0.20 M Htta). The scattering experiments confirm the EXAFS results. The best fit of the scattering data was an Eu inner coordination sphere composed of 7.5 ± 0.6 oxygen atoms at an average Eu-O distance of 2.35 Å. Fitting the experimental scattering data to the results of molecular dynamics simulations of Eu complexes in $C_4mim^+Tf_2N^-$ gave the best results for Eu surrounded by four bidentate tta⁻ anions. Complexes with other compositions, for example $Eu(tta)_3(H_2O)_2$, could not reproduce the experimental scattering unless some of the fitting parameters had unreasonable values. Our experimentally measured Eu-O bond distances are considerably shorter than the average Eu-O distance reported for crystalline $Eu(tta)_3(H_2O)_2$, 2.44 Å (23), but they agree well with the reported distances for the 1,4-dimethylpyridinium salt of $Eu(tta)_4^-$, 2.39 Å (24).

The spectroscopic and scattering experiments all point toward the formation of $Ln(tta)_4^-$ anions with no coordinated water molecules under the extraction conditions that favored 1:4 Eu:tta⁻ complexes in the radiotracer experiments (Figure 2). Although RTILs are inherently ionic and polar solvents, there can be no net electrical charge at equilibrium in either phase of a liquid-liquid extraction system. Thus, if $Ln(tta)_4^-$ is present in the RTIL, there must be a countercation present to balance its charge. The cations available to balance the negative charge of $Ln(tta)_4^-$ are Na^+, H^+, and C_4mim^+. Conclusive demonstration of the presence of $Ln(tta)_4^-$ in the RTIL phase depends on determining if there is a counter ion for the suspected $Ln(tta)_4^-$ anion in the RTIL phase. Identification of the countercation also is critical to quantifying the extraction equilibria.

In conjunction with these experiments, we studied the partitioning of the ionic liquid anion, Tf_2N^-, between the RTIL phase and a 1.0 M NaCl aqueous phase as a function of the amount of the $Nd(tta)_4^-$ complex in the RTIL phase. If the $Ln(tta)_4^-$ complex is extracted into the RTIL by an anion exchange mechanism, the extraction of one molecule of $Ln(tta)_4^-$ into $C_4mim^+Tf_2N^-$ should displace one molecule of Tf_2N^- into the aqueous phase. In fact, this is what we observe. The aqueous phase concentration of Tf_2N^- increases linearly as the concentration of $Nd(tta)_4^-$ increases in an RTIL phase containing 0.5 M $Htta/C_4mim^+Tf_2N^-$.

Combining all of this information, the two lanthanide partitioning equilibria operating in $C_4mim^+Tf_2N^-$ can be identified. For Nd^{3+} and Eu^{3+} They are

$$Ln(H_2O)_h^{3+} + 3\,\overline{Htta} \rightleftharpoons \overline{Ln(tta)_3(H_2O)_3} + 3\,H^+ + (h\text{-}3)\,H_2O, \qquad (5)$$

which dominates at $[tta^-] < 1 \times 10^{-6}$ M with $K_{ex} = D_{Ln}[\,\overline{Htta}\,]^3[H^+]^{-3}$, and

$$Ln(H_2O)_h^{3+} + 4\,\overline{Htta} + \overline{C_4mimTf_2N} \rightleftharpoons$$
$$\overline{C_4mim(Ln(tta)_4)} + 4\,H^+ + Tf_2N^- + h\,H_2O, \qquad (6)$$

which is important when $[tta^-] > 1 \times 10^{-6}$ M. For the sake of comparison to literature data on the extraction $Eu(tta)_4^-$ by tetraalkylammonium liquid anion exchangers, Eq. 6 can be rewritten as

$$Ln(H_2O)_h^{3+} + 4\,tta^- + C_4mim^+ \rightleftharpoons \overline{C_4mim(Ln(tta)_4)} + h\,H_2O, \qquad (7)$$

with $K_{ex4,1} = D_{Ln}[tta^-]^{-4}[C_4mim^+]^{-1}$.

Extraction equilibria comparable to Eq. 5 and Eq. 6 have been reported for Ln^{3+} extraction into molecular solvents. The extraction constants for both Htta partitioning (K_d) and for partitioning of the neutral 1:3 Eu^{3+}-tta^- complex (K_{ex}) are known for many organic solvents (15,21,25-33). The K_d and K_{ex} values, without correction for modest differences in the ionic strength of the aqueous phases, are summarized in Figure 4. As described above, the Htta distribution into $C_4mim^+Tf_2N^-$ resembles both that observed for conventional aromatic solvents and for oxygenated solvents (alcohols, esters, or ketones). In contrast, comparison of the extraction constants for the 1:3 Eu:tta complex in molecular solvents and $C_4mim^+Tf_2N^-$ is not ambiguous. While the extraction constant of $Eu(tta)_3(H_2O)_3$ with $C_4mim^+Tf_2N^-$ (Log K_{ex} = -6.03) falls in the same range as the extraction constants of the oxygenated solvents, it is 30-100 times larger than those observed for aliphatic, aromatic, or chlorinated solvents. Extraction of $Eu(tta)_3(H_2O)_3$ into $C_4mim^+Tf_2N^-$ resembles that of oxygenated molecular solvents, not non-polar solvents. This mirrors what is expected from the reported polarity of RTILs (16).

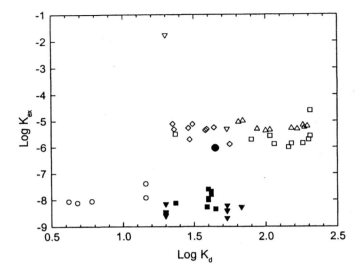

Figure 4. Dependence of the Eu(tta)₃ and Htta extraction constants on organic solvent class: (■) aromatic, (O) aliphatic, (▼) chlorinated, (△) ester, (□) ketone, (◊) alcohol, (▽) 0.1 M tributylphosphate in CCl₄ or CHCl₃, and (●) C₄mim⁺Tf₂N⁻ (15,25,29-37).

The reason for the similarities of Eu(tta)₃(H₂O)₃ extraction into C₄mim⁺Tf₂N⁻ and oxygenated molecular solvents is not clear. One of two interrelated mechanisms could be at work. Both oxygenated organic solvents and RTILs are known to dissolve significant concentrations of water. For example, 1-octanol dissolves between 2 and 3 M H₂O, depending on the acidity of the aqueous phase (*34*). The water content of equilibrated C₄mim⁺Tf₂N⁻ is somewhat less (Table I), but it is still substantial in comparison to the non-polar organic solvents with low K_{ex} values. The solvation energy of a hydrated complex like Eu(tta)₃(H₂O)₃ should be more favorable in a water rich organic solvent than in a solvent that dissolves little water (e.g, benzene or octane). Another possibility is that both the oxygenated molecular solvents and C₄mim⁺Tf₂N⁻ directly interact with the extracted complex, though probably in different ways. Oxygenated molecular solvents are believed to boost the extraction of Ln^{3+}-tta⁻ complexes by replacing one or more of the coordinated water molecules in Ln(tta)₃(H₂O)ₘ, producing a more hydrophobic complex that is more soluble in the organic phase (*32*). However, since the optical absorption spectrum of the Nd^{3+}-bza⁻ complex in $C_4mim^+Tf_2N^-$ is a good match for the spectrum of $Nd(tta)_3(H_2O)_3$, the RTIL probably does not displace coordinated water molecules from the extracted complex by direct coordination to the Ln^{3+}. Instead, if there is a specific interaction between the RTIL components and $Ln(tta)_3(H_2O)_m$, it should take place in the metal's outer coordination sphere, probably through hydrogen bonding between the RTIL components and complex, as has been reported for other extraction systems (35).

The anion exchange of $Ln(tta)_4^-$ complexes into $C_4mim^+Tf_2N^-$ also can be compared to similar anion exchange reactions in conventional organic solvents. Table III compares the extraction constant for Eq. 7 to the formation and extraction of $Ln(tta)_4^-$ into chloroform by anion exchange with the Cl^- or NO_3^- anions of hydrophobic tetraalkylammonium salts. In molecular solvents, the $Ln(tta)_4^-$ extraction increases as the cation becomes more hydrophobic and the anion becomes more hydrophilic (15,36). Similar effects are expected in RTILs, however, using Eq. 7 to compare the extraction removes the anion contribution to the extraction constant, allowing direct comparison of the extraction in sytems with different anions. The extraction of $Ln(tta)_4^-$ into the ionic liquid is comparable to that observed in chloroform containing tetrapropylammonium cations and Cl^- or NO_3^- anions. This result suggests that using a more hydrophobic cation, for example R_4N^+ or a methylimidazolium with a longer 1-alkyl group could increase the likelihood of observing the anion exchange mechanism for a given RTIL. It also underscores the complexity of RTILs as solvents. While the extraction of the neutral 1:3 complex into $C_4mim^+Tf_2N^-$ resembles that of oxygenated molecular solvents (Figure 4), the anion exchange mechanism required for the extraction of the 1:4 complex was not observed in oxygenated molecular solvents in the presence of tetraalkylammonium salts (15).

Table III. Extraction constants for the partitioning of Eu(tta)$_4^-$ ion pairs into CHCl$_3$ by anion exchange for Cl$^-$ or NO$_3^-$ (37) or into C$_4$mim$^+$Tf$_2$N$^-$ by anion exchange for Tf$_2$N$^-$

Cation	Log $K_{ex4,1}$ when the anion is		
	NO_3^-	Cl^-	Tf_2N^-
$(C_3H_7)_4N^+$	24.9	25.0	
$(C_4H_9)_4N^+$	27.1	27.1	
$(C_5H_{11})_4N^+$	29.4	29.5	
$(C_{14}H_{29})(CH_3)_2(C_6H_5CH_2)N^+$	32.1	32.0	
C_4mim^+			25.1 ± 0.1

Conclusions

The behavior of solutes in biphasic water-RTIL systems is varied and complex, and is not always well represented by a single parameter. Despite this, the complexes of trivalent lanthanide cations with Htta that form in $C_4mim^+Tf_2N^-$ are directly comparable to the complexes formed in molecular organic solvents, and they partition between the phases by related extraction mechanisms. The affinity of the neutral complex $Ln(tta)_3(H_2O)_m$ for $C_4mim^+Tf_2N^-$ is similar to that of polar, oxygenated molecular solvents (alcohols, esters, ketones), and is 1-2 orders of magnitude greater than for non-polar aliphatic, aromatic, or chlorinated solvents. The ionic nature of the RTIL phase also allows the formation of the anionic complex, $Ln(tta)_4^-$, which previously has only been observed in low polarity molecular organic solvents (alkane, aromatic, or chlorinated solvents) in the presence of hydrophobic countercations like tetraalkylammonium salts. In this case, the extraction of $Ln(tta)_4^-$ (Eq. 6) is opposed by the less favorable hydration energy of Tf_2N^- as compared to literature examples, which exchanged Cl^- or NO_3^- for $Ln(tta)_4^-$, but its extraction into RTILs might be enhanced by substituting more hydrophobic RTIL cations for C_4mim^+.

Acknowledgments

P. Rickert purified the Htta and S. Naik determined the RTIL H_2O content. BESSRC-CAT and the Actinide Facility provided infrastructure for the X-ray work. This work was supported by the U.S.D.O.E., Office of Basic Energy Science through contract number W-31-109-ENG-38.

Literature Cited

1.   Huddleston, J. G.; Willauer, H. D.; Swatloski, R. P.; Visser, A. E.; Rogers, R. D. *Chem. Commun.* **1998**, 1765.
2.   Laintz, K. E.; Wai, C. M.; Yonker, C. R.; Smith, R. D. *Anal. Chem.* **1992**, *64*, 2875.
3.   Rogers, R. D.; Bond, A. H.; Griffin, S. T.; Horwitz, E. P. *Solvent Extr. Ion Exch.* **1996**, *14*, 919.
4.   Visser, A. E.; Holbrey, J. D.; Rogers, R. D. *Chem. Commun.* **2001**, 2484.
5.   Dietz, M. L.; Dzielawa, J. A. *Chem. Commun.* **2001**, 2124.
6.   Visser, A. E.; Jensen, M. P.; Laszak, I.; Nash, K. L.; Choppin, G. R.; Rogers, R. D. *Inorg. Chem.* **2003**, *42*, 2197.

7. Jensen, M. P.; Dzielawa, J. A.; Rickert, P.; Dietz, M. L. *J. Am. Chem. Soc.* **2002**, *124*, 10664.
8. Visser, A. E.; Swatloski, R. P.; Reichert, W. M.; Griffin, S. T.; Rogers, R. D. *Ind. Eng. Chem. Res.* **2000**, *39*, 3596.
9. Dai, S.; Ju, Y. H.; Barnes, C. E. *J. Chem. Soc., Dalton Trans.* **1999**, 1201.
10. Chen, S.; Dzyuba, S.; Bartsch, R. A. *Anal. Chem.* **2001**, *73*, 3737.
11. Hasegawa, Y.; Ishiwata, E.; Ohnishi, T.; Choppin, G. R. *Anal. Chem.* **1999**, *71*, 5060.
12. Mathur, J. N.; Choppin, G. R. *Solvent Extr. Ion Exch.* **1993**, *11*, 1.
13. Kitatsuji, Y.; Meguro, Y.; Yoshida, Z.; Yamamoto, T.; Nishizawa, K. *Solvent Extr. Ion Exch.* **1995**, *13*, 289.
14. Kononenko, L. I.; Vitkun, R. A. *Russ. J. Inorg. Chem.* **1970**, *15*, 1345.
15. Noro, J.; Sekine, T. *Bull. Chem. Soc. Jap.* **1993**, *66*, 450.
16. Aki, S. N. V. K.; Brennecke, J. F.; Samanta, A. *Chem. Commun.* **2001**, 413.
17. Jensen, M. P.; Beitz, J. V.; Neuefeind, J.; Skanthakumar, S.; Soderholm, L. *J. Am. Chem. Soc.* **2003**, *125*, 15466.
18. Sekine, T.; Hasegawa, Y.; Ihara, N. *J. Inorg. Nucl. Chem.* **1973**, *35*, 3968.
19. Suzuki, N.; Akiba, K. *J. Inorg. Nucl. Chem.* **1971**, *33*, 1897.
20. Anderson, J. L.; Ding, J.; Welton, T.; Armstrong, D. W. *J. Am. Chem. Soc.* **2002**, *124*, 14247.
21. Sekine, T.; Dyrssen, D. *J. Inorg. Nucl. Chem.* **1967**, *29*, 1481.
22. Horrocks, W. D., Jr.; Sudnick, D. R. *J. Am. Chem. Soc.* **1979**, *101*, 334.
23. White, J. G. *Inorg. Chim. Acta* **1976**, *16*, 159.
24. Chen, X.-F.; Liu, S.-H.; Duan, C.-Y.; Xu, Y.-H.; You, X.-Z.; Ma, J.; Min, N.-B. *Polyhedron* **1998**, *17*, 1883.
25. Poskanzer, A. M.; Foreman, B. M. *J. Inorg. Nucl. Chem.* **1961**, *16*, 323.
26. Sekine, T.; Dyrssen, D. *J. Inorg. Nucl. Chem.* **1967**, *29*, 1457.
27. Sekine, T.; Ono, M. *Bull. Chem. Soc. Jap.* **1965**, *38*, 2087.
28. Alstad, J.; Augustson, K. H.; Farbu, L. *J. Inorg. Nucl. Chem.* **1974**, *36*, 899.
29. Kassierer, E. F.; Kertes, A. S. *J. Inorg. Nucl. Chem.* **1972**, *34*, 3221.
30. Bhatti, M. S.; Desreux, J. F.; Duyckaerts, G. *J. Inorg. Nucl. Chem.* **1980**, *42*, 767.
31. Akiba, K.; Wada, M.; Kanno, T. *J. Inorg. Nucl. Chem.* **1981**, *43*, 1031.
32. Akiba, K.; Kanno, T. *J. Inorg. Nucl. Chem.* **1980**, *42*, 273.
33. Sekine, T.; Takahashi, Y.; Ihara, N. *Bull. Chem. Soc. Japan* **1973**, *46*, 388.
34. Sun, Y.; Moyer, B. A. *Solvent Extr. Ion Exch.* **1995**, *13*, 243.
35. Kameta, N.; Imura, H.; Ohashi, K.; Aoyama, T. *Polyhedron* **2002**, *21*, 208.
36. Dukov, I. L.; Atanassova, M. *Hydrometallurgy* **2003**, *68*, 89.
37. Noro, J.; Sekine, T. *Bull. Chem. Soc. Japan* **1993**, *66*, 804.

Chapter 3

An EXAFS, X-ray Diffraction, and Electrochemical Investigation of 1-Alkyl-3-methylimidazolium Salts of [{UO$_2$(NO$_3$)$_2$}$_2$(μ$_4$-C$_2$O$_4$)]$^{2-}$

Antonia E. Bradley[1], Christopher Hardacre[1],
Mark Nieuwenhuyzen[1], William R. Pitner[1], David Sanders[1],
Kenneth R. Seddon[1], and Robert C. Thied[2]

[1]The QUILL Centre and The School of Chemistry, Queen's University
of Belfast, Belfast BT9 5AG, United Kingdom
[2]British Nuclear Fuels plc, B170, Sellafield, Seascale, Cumbria CA20 1PG,
United Kingdom

The structure of the 1-alkyl-3-methylimidazolium salts of the dinuclear μ$_4$-(*O,O,O',O'*-ethane-1,2-dioato)-bis[bis(nitrato-*O,O*)dioxouranate(VI)] anion have been investigated using single crystal X-ray crystallography. In addition, EXAFS and electrochemical studies have been performed on the [C$_4$mim]$^+$ salt which is formed following the oxidative dissolution of uranium(IV) oxide in [C$_4$mim][NO$_3$]. EXAFS analysis of the solution following UO$_2$ dissolution indicates a mixture of uranyl nitrate and μ$_4$-(*O,O,O',O'*-ethane-1,2-dioato)-bis[bis(nitrato-*O,O*)dioxouranate(VI)] anions are formed.

Introduction

An accelerating number of publications demonstrate the growing interest in the use of ionic liquids by the nuclear industry (*1-7*). Beginning with the work of D'Olieslager and coworkers (*1*) and Hitchcock *et al.* (*2a*) the electrochemical and spectrochemical behaviour of dioxouranium(VI) species in chloroaluminate room-temperature ionic liquids have been investigated (*2*). Costa *et al.* (*2e*) recently compared the dioxouranate(VI) and dioxoplutonate(VI) chemical and electrochemical systems in acidic chloroaluminate ionic liquids, and presented arguments for the potential use of ionic liquids throughout the nuclear industry. It has also been demonstrated that room temperature ionic liquids can be used for solvent extraction of metal species from aqueous media (*3*). This is an area of great significance to the nuclear industry, which currently uses solvent extraction in the PUREX process for reprocessing spent nuclear fuel (*4*). The chemistry of lanthanides in ionic liquids, including the electrodeposition of these metals, has been the subject of a number of papers (*5*). Critical mass calculations performed by Harmon *et al.* on two plutonium metal/ionic liquid mixtures have recently been published (*6*). Recent work from our laboratories has led to the publication of a series of patents (*7*) and papers (*8*) concerned with using ionic liquids in nuclear fuel reprocessing and molten salt waste treatment.

In this paper we examine the formation and structure of μ_4-(O,O,O',O'-ethane-1,2-dioato)-bis[bis(nitrato-O,O)dioxouranate(VI)]$^{2-}$ based 1-alkyl-3-methylimidazolium salts. These salts have been shown to be formed on the oxidative dissolution of UO_2 in nitrate based ionic liquids (*8a*). The paper also describe an EXAFS study solutions following the oxidative dissolution of UO_2 in [C_4mim][NO_3] in the presence and absence of acetone in order to investigate the origin of the oxalate species. We also describe an attempt to isolate the uranium, following oxidative dissolution of UO_2 in [C_4mim][NO_3], through electrochemical reduction to an oxouranium(IV) species.

Experimental

Sample Preparation.

The 1-alkyl-3-methylimidazolium nitrates [$C_n$mim][NO_3] (n = 4, 6, 16) were prepared by metathesis reactions between silver(I) nitrate and the appropriate [$C_n$mim]Cl, carried out in water. The silver(I) chloride precipitate was removed by filtration from the aqueous mixture and the water was removed from the [$C_n$mim][NO_3] by evaporation under reduced pressure. To remove as much residual silver chloride as possible, the [$C_n$mim][NO_3] was repeatedly dissolved in ethanenitrile and mixed with decolourising charcoal, which acts as an effective nucleation site for the precipitation of silver chloride. The charcoal was removed by filtration and the ethanenitrile was removed by evaporation under reduced pressure.

The preparation of [$C_n$mim]$_2$[{$UO_2(NO_3)_2$}$_2$(μ-C_2O_4)] involved mixing [$C_n$mim][NO_3] (10 g), uranium(IV) nitrate hexahydrate (1 g), concentrated nitric acid (1 g) and acetone (0.1 g). This mixture was heated to 70 °C for 2 h and allowed to cool. For $n = 4$ and 6, this procedure was found to result in the precipitation of a yellow solid which could be separated by filtration and recrystallised from ethanenitrile. For $n = 16$, cooling did not cause precipitation, and the formation of the solid salt was achieved by dissolving the reaction mixture in a minimum amount of ethanenitrile, addition of ethyl ethanoate, and cooling of the mixture.

Single Crystal X-ray Crystallography.

Experimental parameters and crystal data are listed in Table 1. Data were collected on a Siemens P4 diffractometer using the XSCANS software with omega scans for [C_4mim]$_2$[{$UO_2(NO_3)_2$}$_2$(μ-C_2O_4)]. For all other structures, a Bruker SMART diffractometer using the SAINT-NT software was used. A crystal was mounted on to the diffractometer at low temperature under dinitrogen at *ca.* 120 K. The structures were solved using direct methods with the SHELXTL program package and the non-hydrogen atoms were refined with anisotropic thermal parameters. Hydrogen-atom positions were added at idealized positions and refined using a riding model. Lorentz, polarisation and empirical absorption corrections were applied. The function minimized was $\Sigma[w(|F_o|^2 - |F_c|^2)]$ with reflection weights $w^{-1} = [\sigma^2 |F_o|^2 + (g_1P)^2 + (g_2P)]$ where $P = [\max |F_o|^2 + 2|F_c|^2]/3$. Additional material available from the Cambridge Crystallographic Data Centre comprises relevant tables of atomic coordinates, bond lengths and angles, and thermal parameters (CCDC Nos. 224134, 224136). [C_4mim]$_2$[{$UO_2(NO_3)_2$}$_2$(μ-C_2O_4)] and [C_{16}mim]$_2$[{$UO_2(NO_3)_2$}$_2$(μ-C_2O_4)] were prepared for X-ray crystallography by recrystallisation of the precipitate from ethanenitrile. [C_6mim]$_2$[{$UO_2(NO_3)_2$}$_2$(μ-C_2O_4)] was obtained directly from the ionic liquid.

Extended X-ray Absorbance Fine Structure (EXAFS) Spectroscopy.

Data were collected at the Synchrotron Radiation Source in Daresbury, U.K., using station 9.2. The transmission detection mode was used with two ionization chambers filled with argon. The spectra were recorded at the uranium L_{III} edge using a double crystal Si(220) monochromator set at 50 % harmonic rejection. A minimum of three scans per sample were required to obtain sufficiently good signal to noise. Scans were collected and averaged using EXCALIB which was also used to convert raw data into energy vs. absorption data. EXBROOK was used to remove the background. The analysis of the EXAFS was performed using EXCURV98 on the raw data using the curved wave theory. Multiple scattering was employed for the dioxouranium(VI) oxygens and the oxalate groups defined as individual units within the analysis package. The solution data was measured using procedures described in detail elsewhere (9). Data was analysed up to $k = 13$ Å^{-1} as k^3 weighted data with an value of $AFAC = 1.0$ and the with an $R_{max} = 10$ Å.

Electrochemical Analysis.

All electrochemical experiments were carried out with an EG&G PARC Model 283 potentiostat/galvanostat connected to a PC through an IEEE-488 bus and controlled using EG&G Parc Model 270/250 Research Electrochemistry version 4.23 software. Positive feedback iR compensation was employed to eliminate errors due to solution resistance. The electrochemical cell was constructed from materials purchased from Bioanalytical Systems, Inc. (BAS). The non-aqueous reference electrode was a silver wire immersed in a glass tube containing a 0.100 mol l^{-1} solution of AgNO$_3$ in the [C$_n$mim][NO$_3$] ionic liquid separated from the bulk solution by a Vycor plug. All potentials reported are referenced against the AgNO$_3$/Ag couple. The counter electrode was a platinum coil immersed directly in the bulk solution. The working electrode was a glassy carbon disk ($A = 7.07 \times 10^{-2}$ cm^2). The solution was held in a glass vial fitted with a Teflon cap with holes for the electrodes and a gas line. All work was carried out under a dry dinitrogen atmosphere, and dinitrogen was bubbled through the solution prior to performing electrochemical experiments. All electrochemical experiments were carried out at 40 °C with a scan rate of 50 mV s^{-1}.

Table 1 X-ray experimental crystal data.

	1	2*	3
Empirical Formula	$C_{18}H_{30}N_8O_{20}U_2$	$C_{32}H_{57}N_{11}O_{23}U_2$	$C_{52}H_{93}N_{13}O_{20}U_2$
Formula weight 30 g mol⁻¹	1154.56	1439.95	1696.45
Space group	$P2_1/n$	$P2_1/c$	P-1
Unit cell dimensions			
a Å - α	15.452(2), 90	14.722(3), 90	9.149(2), 94.643(4)
b Å - β	20.354(3), 106.840(15)	32631(7), 96.555(5)	9.973(2), 96.437(4)
c Å - γ	10.822(4), 90	10.587(2), 90	24.330(6), 113.847(4)
Volume Å³, Z	3257.6(13), 4	5052.5(19), 4	1998.1(8), 1
Density (calc.) Mg m⁻³	2.354	1.893	1.410
Crystal dimensions (mm)	0.51 x 0.33 x 0.09	0.20 x 0.10 x 0.06	0.52 x 0.20 x 0.16
Absorption coeff. mm⁻¹	10.023	6.489	4.112
F(000)	2152	2776	840
θ range for data collection	2-45	1.25-22.5	0.85-25
No. Reflections collected	5100	39166	16142
Independent reflec. (R_{int})	4033	6604 (0.1723)	7012 (0.1166)
Final R indices R1 (wR2)	0.0584 (0.1459)	0.1506 (0.4463)	0.0937 (0.2840)

*Crystals were small and of a poor quality but connectivity has been established. The structure has been included because of the presence of an ionic liquid as a 'solvent' of crystallisation.

SOURCE: Reproduced from *Inorg. Chem.* **2004**, *43(8)*, 2503–2514. Copyright 2004 American Chemical Society.

Results

X-ray Crystallography.

1-Butyl-3-methylimidazolium μ₄-(O,O,O',O'-ethane-1,2-dioato)-bis[bis(nitrato-O,O)-dioxouranate(VI)] (1)

X-ray analysis of **1** shows the unit cell contains four $[C_4mim]^+$ cations and two independent $[\{UO_2(NO_3)_2\}_2(\mu\text{-}C_2O_4)]^{2-}$ anions both, of which are located about inversion centres (Figure 1). The $[C_4mim]^+$ cations are arranged such that they produce large channels in which the anions are located. Thus, the anions effectively act as a template for the cations. The flexibility of the alkyl chains on the cations is illustrated by the differences in the C2-N1-C6-C7 torsion angles for the butyl chain at 104.7° and 129.1°.

1-Hexyl-3-methylimidazolium μ₄-(O,O,O',O'-ethane-1,2-dioato)-
bis[bis(nitrato-O,O)-dioxouranate(VI)]: 1-hexyl-3-methylimidazolium nitrate
(1/1) (2)

The asymmetric unit contains one oxalate-bridged dinuclear anion, one nitrate anion and three unique cations, one of which is disordered (Figure 2). The dimers are aligned in columns in the (001) direction with adjacent columns being parallel to each other. The anions still act as templates for the cation, but because of the length of the hexyl chains the structure has crystallised as a solvate containing the 1-hexyl-3-methylimidazolium nitrate ionic liquid. The general features of the structure are similar to **1**.

The structure of **2** is unusual in that it is a co-crystal of the $[C_6mim]_2[\{UO_2(NO_3)_2\}_2(\mu_4-C_2O_4)]$ and the $[C_6mim][NO_3]$ ionic liquid. The incorporation of ionic liquids in crystal lattices is very rare and there is only one other known example (*10*). The co-crystallisation is probably due to the length of the alkyl chains on the cations. These are not long enough to be able to form an interdigitated structure as found in the 1-hexadecyl-3-methylimidazolium salt, **3**, below. To overcome this, the nitrate ionic liquid solvent cocrystallises with the dioxouranium(VI) salt.

1-Hexadecyl-3-methylimidazolium μ₄-(O,O,O',O'-ethane-1,2-dioato)-
bis[bis(nitrato-O,O)-dioxouranate(VI)]:ethanenitrile (1/6) (3)

The asymmetric unit contains one unique anion which is located about an inversion centre and one cation and three ethanenitrile molecules, two of which are disordered. The cations and anions are arranged such that the anions are associated in columns along the (001) direction with the imidazolium head groups hydrogen bonding to these columns. The alkyl chains extend away from these 'charged' regions, forming an interdigitated bilayer. This arrangement leads to large channel-like 'voids' in this lattice, with the closest contact between alkyl chains being 8.0 Å, Figure 3. The voids are filled with ethanenitrile solvent molecules, which are disordered and thus adopt random positions throughout the channels.

EXAFS

Figure 4 shows the *pseudo* radial distribution functions obtained from solutions of $UO_2(NO_3)_2.6H_2O$ and **1** dissolved in $[C_4mim][NO_3]$ at room temperature and the reaction solution following the oxidative dissolution of UO_2 in $[C_4mim][NO_3]$, without added acetone and in the presence of 0.2 M

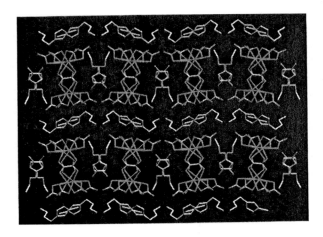

*Figure 1. Illustration of the cation-anion arrangement in **1** shows the cations surrounding the anion columns. Hydrogen atoms have been omitted for clarity.* (Reproduced from *Inorg. Chem.* **2004**, *43(8)*, 2503-2514. Copyright 2004 American Chemical Society.) *(See page 1 of color insert.)*

*Figure 2. Illustration of the cation-anion arrangement in **2** shows the cations and solvate ionic liquid surrounding the anion columns. Hydrogen atoms have been omitted for clarity.* (Reproduced from *Inorg. Chem.* **2004**, *43(8)*, 2503-2514. Copyright 2004 American Chemical Society.) *(See page 1 of color insert.)*

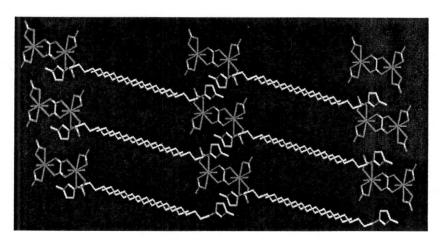

Figure 3. Illustration of the cation-anion arrangement in 3 showing the anions associated with cationic head groups and channels between the alkyl chains regions. The MeCN molecules have been omitted for clarity.
(Reproduced from *Inorg. Chem.* **2004**, *43(8)*, 2503-2514. Copyright 2004 American Chemical Society.) *(See page 2 of color insert.)*

acetone. The parameters used to model the experimental data are summarised in Table 2. On dissolution of 1 in [C$_4$mim][NO$_3$], the EXAFS indicates that the oxalate "dinuclear" species remains intact in solution and the hydrated dioxouranium(VI) cation is not formed. The analysis also shows that the η^2-NO$_3$ are not displaced by water. This may be compared with the *in-situ* formation of dioxouranium(VI) oxalate species in water. Vallet *et al.* showed that on dissolution of uranyl nitrate in the presence of sodium oxalate, [UO$_2$(oxalate)$_2$(H$_2$O)]$^{2-}$ was formed (*11*). There are two main multiple scattering pathways that contribute significantly to the final fit for the EXAFS data: one derived from the dioxouranium(VI) unit and the other from the oxalate unit, in agreement with other studies (*11,12*). The U-U distance present in 1 is not seen in the EXAFS data due to the long length of the U...U distance, 6.3 Å. Where peaks have been noted in other dinuclear structures, the U...U distances are below 4 Å (*12*).

Comparing the EXAFS data for the pure dinuclear oxalate and mononuclear nitrate complexes dissolved in the ionic liquid with the solutions obtained after dissolution of UO$_2$ in [C$_4$mim][NO$_3$] indicates that a mixture of species is present. Furthermore, the addition of acetone had little effect on the EXAFS data. For both solutions, following dissolution of UO$_2$, the best fit was

40

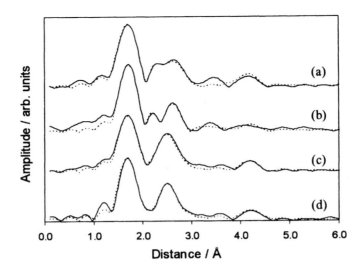

*Figure 4 Comparison of the experimental (solid line) and fitted (dashed line)
EXAFS (upper) and pseudo-radial distribution (lower) functions from (a)
UO₂(NO₃)₂.6H₂O and (b) [C₄mim]₂[{UO₂(NO₃)₂}₂(μ₄-C₂O₄)], 1, dissolved in
[C₄mim][NO₃] and following UO₂ dissolution in [C₄mim][NO₃] (c) in the
absence of acetone and (d) in the presence of 0.2 M acetone.*
(Reproduced from *Inorg. Chem.* **2004**, *43(8)*, 2503-2514. Copyright 2004
American Chemical Society.)

obtained with 15:85 mononuclear nitrate: dinuclear oxalate mole ratio present
in solution. The error in the proportion of nitrate monomer present is high (±
10 mol %) due to the shallowness of the minimum found in the fit on varying
the composition.

The mixture postulated from the EXAFS results is consistent with UV-vis
and IR measurements following oxidative dissolution of uranium(IV) oxide in
the ionic liquid. The UV-vis spectrum of the yellow solution confirmed that
uranium(VI) was present but the local environment of the species could only be
postulated as a monomeric dioxouranium(VI) species with nitrate ligands.
Infra-red spectra of the liquid showed an identical fingerprint pattern in the
1000 – 1500 cm⁻¹ region to that of **1**, which is assigned to tetradentate oxalate
C-O stretching.). It should be noted that it is not possible to determine
whether the dimer is still present or whether a mononuclear species has been
formed. For example, the fact that Debye-Waller factors for the U-C distances
are so high may suggest that the oxalate species is μ₂-bonded.

Although good agreement between the EXAFS and the hydrated
dioxouranium(VI) nitrate species is obtained, there are a number of other
possible structural arrangements on dissolution of UO₂(NO₃)₂.6H₂O in the
nitrate ionic liquid involving monodentate nitrate co-ordination, namely

Table 2 Structural parameters from the fitted EXAFS spectra from $UO_2(NO_3)_2.6H_2O$ and $[C_4mim]_2[\{UO_2(NO_3)_2\}_2(\mu_4-C_2O_4)]$, 1, dissolved in $[C_4mim][NO_3]$ and following UO_2 dissolution in $[C_4mim][NO_3]$ in the absence of acetone and in the presence of 0.2 M acetone. (Ox) denotes oxalate, (U=O) denotes axial dioxouranium(VI) oxygens.

Solution	Atom	$R / Å$	Co-ord. Number	$\sigma^2 / Å^2$	Fit / %
$UO_2(NO_3)_2.6H_2O/$ [C_4mim][NO_3]	O (U=O)	1.75	2.0	0.005	18.1
	O (H$_2$O)	2.34	2.0	0.014	
	O (NO$_3$)	2.51	4.0	0.013	
	N (NO$_3$)	2.96	2.0	0.005	
1/[C_4mim][NO_3]	O (U=O)	1.76	2.0	0.004	25.6
	O (NO$_3$)	2.52	4.0	0.014	
	N (NO$_3$)	2.89	2.0	0.001	
	O (Ox)	2.32	2.0	0.009	
	C (Ox)	3.06	2.0	0.001	
$UO_2/$ [C_4mim][NO_3] without acetone	O (U=O)	1.75	2.0	0.007	19.2
	O (H$_2$O)	2.33	0.3	0.017	
	O (NO$_3$)	2.49	4.0	0.009	
	N (NO$_3$)	2.95	2.0	0.006	
	O (NO$_3$)	4.15	2.0	0.013	
	O (Ox)	2.33	1.7	0.019	
	C (Ox)	3.23	1.7	0.050	
	O (Ox)	4.53	1.7	0.015	
$UO_2/$ [C_4mim][NO_3] with 0.2 M acetone	O (U=O)	1.75	2.0	0.006	19.2
	O (H$_2$O)	2.36	0.3	0.026	
	O (NO$_3$)	2.47	4.0	0.007	
	N (NO$_3$)	2.95	2.0	0.005	
	O (NO$_3$)	4.15	2.0	0.013	
	O (Ox)	2.28	1.7	0.011	
	C (Ox)	3.28	1.7	0.050	
	O (Ox)	4.55	1.7	0.005	

SOURCE: Reproduced from *Inorg. Chem.* **2004**, *43(8)*, 2503–2514. Copyright 2004 American Chemical Society.

$[UO_2(\eta^2\text{-}NO_3)_2(\eta^1\text{-}NO_3)(H_2O)]^-$ and $[UO_2(\eta^2\text{-}NO_3)_2(\eta^1\text{-}NO_3)_2]^{2-}$. Fitting the EXAFS using these species showed similar fit factors and χ^2 values for both the uranyl nitrate dissolved in [C₄mim][NO₃] and for UO_2 oxidatively dissolved in [C₄mim][NO₃] in the presence and absence of acetone. It is likely that a range of neutral and anionic species is present in each case, and that the solution contains an equilibrium mixture.

Electrochemistry.

The electrochemical behaviour of **1** in [C₄mim][NO₃] was investigated using cyclic voltammetry. A typical cyclic voltammogram is shown in Figure 5. The voltammogram is characterised by a broad irreversible reduction wave with a peak around -0.9 V. The broad oxidation wave with a peak around -0.1 V is associated with the reduction wave, but its relatively small amplitude indicates the irreversible nature of the reduction.

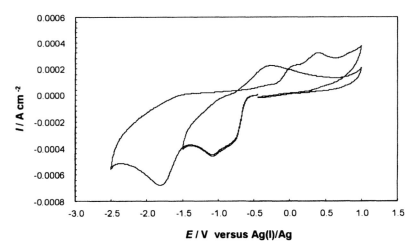

Figure 5 Cyclic voltammogram (scan rate = 50 mV s⁻¹) of
[C₄mim]₂[{UO₂(NO₃)₂}₂(μ₄-C₂O₄)] in [C₄mim][NO₃], recorded at a glassy carbon electrode at 40 °C
(Reproduced from *Inorg. Chem.* **2004**, *43(8)*, 2503-2514. Copyright 2004 American Chemical Society.)

Exhaustive electrolysis was repeatedly carried out at a glassy carbon flag working electrode ($A \sim 1$ cm^2) in an attempt to isolate the product. For constant potential electrolysis, the potential of working electrode was held at suitable negative potential ($\sim$ -1.2 V) to bring about reduction of the [C₄mim]₂[{UO₂(NO₃)₂}₂(μ₄-C₂O₄)]. This led to passivation of the electrode by a thin film of a brown powder. To prevent passivation of the electrode surface, electrolysis was carried out at -1.2 V while periodically pulsing the potential to +1 V using a square-wave potential program. A typical potential program used during electrolysis consisted of holding the potential at -1.2 V for 59.9 s and at +1 V for 0.1 s in a continuous loop. Over time, the electrolysis solution changed from bright yellow to dark brown. Various attempts to isolate a product by filtration and solvent extraction were unsuccessful. However, the electrochemical reduction of **1** is a two-electron metal-centered reduction of uranium(VI) to uranium(IV), which is consistent with formation of a UO_2-type species. Over time, this leads to the electrode being coated with a passivating film of UO_2, which is an insulator at room temperature, and thereafter the current drops to zero.

Discussion

The source of the bridging oxalate moiety in the precipitate formed in the UO_2 oxidative dissolution was suspected to be an organic species inadvertently coming into contact with the reaction mixture. Acetone is a common organic solvent used in the preparation of ionic liquids and the cleaning of glassware and may be oxidised by nitric acid, and therefore was one possible oxalate source. Indeed studies in the nineteenth century showed that acetone and other aliphatic ketones and aldehydes could be oxidised by nitric acid to give oxalic acid (*13*). In fact, in the complete absence of acetone in the ionic liquid, precipitation does not occur.

^{13}C NMR experiments using $(CH_3)_2{}^{13}CO$ or $({}^{13}CH_3)_2CO$ (*ex* Aldrich) showed that acetone could act as a source of oxalte. In the ^{13}C NMR spectra of $[C_4mim]_2[\{(UO_2)(NO_3)_2\}_2(\mu\text{-}C_2O_4)]$, eight peaks are observed from the $[C_4mim]^+$ cation plus a singlet peak at 177 ppm arising from the oxalate carbon atom. However, when the oxidation of uranium(IV) oxide by nitric acid in $[C_4mim][NO_3]$ was carried out in the presence of ^{13}C labelled acetone, the peak at 177 ppm in the ^{13}C NMR spectra showed a significant increased magnitude relative to the other peaks compared with the ^{13}C NMR spectra of $[C_4mim]_2[\{(UO_2)(NO_3)_2\}_2(\mu\text{-}C_2O_4)]$ prepared using unlabelled acetone. It is interesting to note that similar variations were observed whether the acetone was added either prior to or after oxidative dissolution of UO_2 had occurred. Although the NMR data showed incorporation of the acetone into the dinuclear oxalate anion, it is clear from the solution phase EXAFS results that, both in the presence and absence of added acetone, the same species are found in solution and therefore acetone is not the only source of oxalate.

Rogers *et al.* (*14*) reported the preparation of the anion $[\{UO_2(NO_3)_2\}_2(\mu\text{-}C_2O_4)]^-$ by heating dioxouranium(VI) sulfate trihydrate with concentrated nitric acid and a crown ether. The oxalate species was attributed to impurities in the nitric acid, as a variety of acids was used and the bridging oxalate anion was only noted when the complexes were prepared with nitric acid. Glyoxal is a starting material in the preparation of imidazole, which is required for the synthesis of imidazolium-based ionic liquids, and it has been noted that the oxidation of alcohols with concentrated nitric acid results in a variety of species including oxalate (*13*). NMR experiments replacing acetone with glyoxal confirmed that this could also be a source of the oxalate species. It is therefore possible that organic impurities in either the nitric acid or ionic liquid are oxidised to the oxalate ligand, which is trapped between two dioxouranium(VI) nitrate units.

As outlined in the introduction, part of the motivation behind this study was to examine whether the oxidative dissolution of UO_2 in ionic liquids could be used in the processing of spent fuels. A number of electrochemical processes have proposed including the Argonne National Laboratory (ANL) lithium process, where declad oxide fuel is dissolved in a LiCl:KCl eutectic at 773 K. On reduction to the metal, the uranium is then purified through an electrorefining process where the fuel is oxidised at an anode and purified uranium is deposited at a cathode. The Research Institute of Atomic Reactors (RIAR) in Dimitrovgrad, Russia, operates a similar process using a NaCl:KCl melt at 1000 K. The electrochemical processes require the high temperatures to ensure that the UO_2 formed is a conductor and does not passivate the electrode. This is the major disadvantage of the low temperature ionic liquid process described herein, as, without the constant refreshing of the electrode surface to remove the passivating film of UO_2, the current decreases to zero and the process stalls.

Acknowledgements

We thank BNFL for financial support (A.E.B, W.R.P. and D.S.), and the EPSRC and Royal Academy of Engineering for the Award of a Clean Technology Fellowship (K.R.S). The CLRC are acknowledged for the award of the beamtime.

References

1   (a) Dewaele, R.; Heerman, L.; D'Olieslager, W. *J. Electroanal. Chem.* **1982**, *142*, 137-146. (b) Heerman, L.; Dewaele, R.; D'Olieslager, W. *J. Electroanal. Chem.* **1985**, *193*, 289-294.

2   (a) Hitchcock, P.B.; Mohammed, T.J.; Seddon, K.R.; Zora, J.A.; Hussey, C.L.; Ward, E.H. *Inorg. Chim. Acta.* **1986**, *113*, L25-L26. (b) Dai, S.; Toth, L.M.; Hayes, G.R.; Peterson, J.R. *Inorg. Chim. Acta* **1997**, *256*, 143-145. (c) Dai, S.; Shin, Y.S.; Toth, L.M.; Barnes, C.E. *Inorg. Chem.* **1997**, *36*, 4900-4902. (d) Anderson, C.J.; Choppin, G.R.; Pruett, D.J.; Costa, D.; Smith, W. *Radiochimica Acta* **1999**, *84*, 31-36. (e) Costa, D.A.; Smith, W.H.; Dewey, H.J. In *Molten Salts XII: Proceedings of the International Symposium*; Trulove, P.C., De Long, H.C., Stafford, G.R., Deki, S., Eds.; The Electrochemical Society: Pennington, NJ, **2000**, 80. (f) Hopkins, T. A.; Berg, J. M.; Costa, D. A.; Smith, W. H.; Dewey, H. J. *Inorg. Chem.* **2001**, *40*, 1820-1825. (g) Oldham, W. J.; Costa, D. A.; Smith, W.H. In *Ionic Liquids as Green Solvents: Progress and Prospects*; Rogers, R. D., Seddon, K. R., Eds; ACS Symposium Series 818; American Chemical Society: Washington, DC, **2002**, p. 188-198.

3   (a) Dai, S.; Ju, Y.H.; Barnes, C.E. *J. Chem. Soc., Dalton Trans.* **1999**, 1201-1202. (b) Visser, A.E.; Swatloski, R.P.; Reichert, W.M.; Griffin S.T.; Rogers, R.D. *Ind. Eng. Chem. Res.* **2000**, *29*, 3596-3604. (c) Visser, A.E.; Swatloski, R.P.; Reichert, W.M.; Mayton, R.; Sheff, S.; Wiezbicki, A.; Davis, J.H.; Rogers, R.D. *Chem. Commun.* **2001**, 135-136.

4   Naylor, A.; Wilson, P.D. In *Handbook of Solvent Extraction*, Lo, T.C., Baird, M.H.I., Hanson, C., Eds.; John Wiley & Sons: New York, USA, 1983, p.783.

5   (a) Gau, W. J.; Sun, I. W. *J. Electrochem. Soc.* **1996**, *143*, 170-174. (b) Gau, W. J.; Sun, I. W. *J. Electrochem. Soc.* **1996**, *143*, 914-919. (c) Tsuda, T; Nohira, T. Ito, Y. *Electrochimica Acta* **2001**, *46*, 1891-1897. (d) Chun, S; Dzyuba, S. V.; Bartsch, R. A. *Anal. Chem.* **2001**, *73*, 3737-3741. (e) Matsumoto, K.; Tsuda, T.; Nohira, T.; Hagiwara, R; Ito, Y.; Tamada, O. *Acta Crystallogr. Sect. C*, **2002**, *8*, m186-m187. (f) Tsuda, T.; Nohira, T.; Ito, Y. *Electrochim. Acta* **2002**, *47*, 2817-2822. (g) Chaumont, A.; Wipff, G. *Phys. Chem. Chem. Phys.*, **2003**, *5*, 3481-3488.

6   Harmon, C.D.; Smith, W.H.; Costa, D.A. *Radiat. Phys. Chem.*, **2001**, *60*, 157-159.

7   (a) Jeapes, A.J.; Thied, R.C.; Seddon, K.R.; Pitner, W.R.; Rooney, D.W.; Hatter, J.E.; Welton, T. World Pat. WO115175, **2001**. (b) Thied, R.C.; Hatter, J.E.; Seddon, K.R.; Pitner, Rooney D.W.; Hebditch, D. World Pat. WO113379, **2001**. (c) Fields, M.; Thied, R.C.; Seddon, K.R.; Pitner, W.R.; Rooney, D.W. World Pat. WO9914160, **1999**. (d) Thied, R.C.; Seddon, K.R.; Pitner, W.R.; Rooney D.W. World Patent WO9941752, **1999**. (e) Fields, M.; Hutson, G.V. Seddon, K.R.; Gordon, C.M. World Pat. WO9806106, **1998**.

8   (a) Bradley, A. E.; Hatter, J. E.; Nieuwenhuyzen, M.; Pitner, W. R.; Seddon, K. R.; Thied, R.C. *Inorg.Chem.* **2002**, *41*, 1692-1694. (b) Allen, D.; Baston, G.; Bradley, A. E.; Gorman, T.; Haile, A.; Hamblett, I.; Hatter, J. E.; Healey, M. J. F.; Hodgson, B.; Lewin, R.; Lovell, K. V.; Newton, G. W. A.; Pitner, W. R.; Rooney, D. W.; Sanders, D.; Seddon, K. R.; Sims, H.E.; Thied, R. C. *Green Chemistry* **2002**, *4*, 152-158. (c) Pitner, W. R.; Bradley, A. E; Rooney, D. W.; Sanders, D.; Seddon, K. R.; Thied, R. C.; Hatter, J. E. In *Green Industrial Applications of Ionic Liquids*, Rogers, R. D.; Seddon, K. R.; Volkov, S.; Eds.; NATO Science Series II: Vol. 92; Kluwer Academic Publishers: Dordrecht, 2002, pp. 209-226. (d) Baston, G. M. N.; Bradley, A. E.; Gorman, T.; Hamblett, I.; Hardacre, C.; Hatter, J. E.; Healy, M. J. F.; Hodgson, B.; Lewin, R.; Lovell, K. V.; Newton, G. W. A.; Nieuwenhuyzen, M.; Pitner, W. R.; Rooney, D. W.; Sanders, D.; Seddon, K. R.; Simms, H. E.; Thied, R. C. In *Ionic Liquids as Green Solvents: Progress and Prospects*; Rogers, R. D.; Seddon, K. R., Eds; ACS Symposium Series 818; American Chemical Society: Washington, DC, **2002**; pp. 162-167. (e) Bradley, A. E.; Hardacre, C.; Nieuwenhuyzen, M.; Pitner, W. R.; Sanders, D.; Seddon, K. R.; Thied, R.C. *Inorg.Chem.* submitted.

9   Hamill, N.A.; Hardacre, C.; McMath, J. *Green Chem.*, **2002**, *4*, 139-142.

10.   Holbery, J. D.; Nieuwenhuyzen, M.; Seddon, K. R. *unpublished results*.

11 Vallet, V.; Moll, H.; Wahlgren, U.; Szabó, Z.; Grenthe, I. *Inorg. Chem.* **2003**, *42*, 1982-1993.

12 Barnes, C.E.; Shin, Y.; Saengkerdsub, S; Dai, S. *Inorg. Chem.* **2000**, *39*, 862-864.

13 (a) Behrend, R.; Schmitz, J. *Chem. Ber.* **1893**, *26*, 626. (b) Apetz, H.; Hell, C. *Chem. Ber.* **1894**, *27*, 933.

14 Rogers, R. D; Bond, A. H.; Hipple, W. G.; Rollins, A. N.; Henry, R. F. *Inorg. Chem.* **1991**, *30*, 2671-2679.

Chapter 4

Uranium Halide Complexes in Ionic Liquids

Maggel Deetlefs[1], Peter B. Hitchcock[2], Charles L. Hussey[3],
Thamer J. Mohammed[2], Kenneth R. Seddon[1],
Jan-Albert van den Berg[1], and Jalal A. Zora[2]

[1]The QUILL Centre, David Keir Building, Stranmillis Road, The Queen's
University of Belfast, Belfast BT9 5AG, United Kingdom
[2]School of Chemistry and Molecular Sciences, University of Sussex,
Falmer, Brighton BN1 9QJ, United Kingdom
[3]Department of Chemistry and Biochemistry, The University of Mississippi,
University, MS 38677

We report here the syntheses, characterisation and
electrochemistry of some 1-ethyl-3-methylimidazolium,
[emim], uranium halide salts. The electrochemistry of the
uranium halide salts were investigated in both basic and acidic
haloaluminate ionic liquids (ILs). The solid state structures of
the uranium chloride salts have previously been reported, but
have now been re-evaluted using a new statistical model to
determine the presence or absence of weak hydrogen bonding
interactions in the crystalline state.

Introduction

The *rôle* and nature of cation-anion interactions in ambient temperature
ionic liquids have been the subject of much interest, resulting from several
studies in which the importance of hydrogen-bonding interactions, or lack
thereof, in such media has been demonstrated.[1,2,3,4,5] The classically employed
criteria to establish the presence of a hydrogen bond in the solid state were
originally developed by Bondi[6] and involve comparing the experimental sum of

the van der Waals radii of interacting atoms with the sum of the calculated values. Thus, if the experimental distance is shorter than that of the calculated value, it is assumed that the interaction is hydrogen bonded. We have long thought that this cutoff criterion is unsatisfactory. Indeed, previous work by our group [2,7,8] and others[9,10,11] has suggested that the classical definition of the hydrogen bond is an oversimplification and additional factors, *viz.* directional and electrostatic properties must be included when evaluating the presence or absence of real hydrogen bonds in the crystalline state. Very recent work in our group has, for example, shown that hydrogen bonding phenomena occur across a wide range of C-H interaction classes and should not be subjected to the van der Waals radius test in which an arbitrary cutoff value determines the hydrogen bonding character of the interaction involved. Van den Berg and Seddon's work focussed on the directional property of hydrogen bonds, and by developing, and subsequently using, a statistical method to analyse various hydrogen bonds, it was shown "how a myriad of separate molecular aggregates orient hydrogen-containing functionalities in a fashion that approaches a linear orientation".

We have now re-evaluated the solid state structures of $[emim]_2[UCl_6]$ and $[emim]_2[UO_2Cl_4]$, previously published as a communication,[12] to establish the presence or absence of hydrogen bonds in the solid state. We further present the electrochemical properties of $[emim]_2[UX_6]$ and $[emim]_2[UO_2X_4]$ (X = Cl or Br) in a basic[13] $(X(AlCl_3) = 0.444)$ chloroaluminate ionic liquid, as well as the electrochemical properties of $[emim]_2[UBr_6]$ and $[emim]_2[UO_2Br_4]$ in an acidic $(X(AlBr_3) = 0.667)$ tetrabromoaluminate ionic liquid.

Syntheses of 1-Ethyl-3-methylimidazolium Uranium Halide Salts

The syntheses of $[emim]_2[UCl_6]$ and $[emim]_2[UO_2Cl_4]$ have previously been described. The salts, $[emim]_2[UBr_6]$ and $[emim]_2[UO_2Br_4]$, were prepared analogously to their chloride counterparts (Figure 1). Dissolution of either uranium(IV) oxide or uranium metal in concentrated aqueous hydrobromic acid (HBr) affords the $[UBr_6]^{2-}$ dianion, which was then treated with two equivalents of [emim]Br dissolved in concentrated aqueous HBr in the presence or absence of air to give respectively $[emim]_2[UBr_6]$ or $[emim]_2[UO_2Br_4]$ in 73 or 65 % yields.

The basic and acidic haloaluminate ionic liquids were prepared using standard procedures.[14,15]

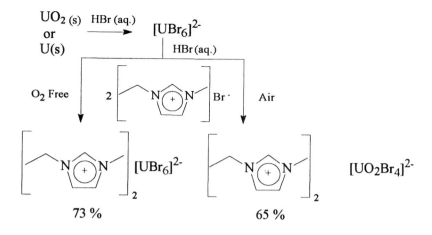

Figure 1. Preparation of [emim]₂[UBr₆] and [emim]₂[UO₂Br₄].

Experimental and Results

Electrochemistry of [emim]₂[UCl₆] and [emim]₂[UO₂Cl₄] in the basic [emim]Cl-AlCl₃ ($X = 0.444$) ionic liquid at 40 °C.

The salt [emim]₂[UCl₆] was dissolved in the basic ionic liquid to produce a light green solution. A cyclic voltammogram of this solution recorded at a scan rate of 50 mV s⁻¹ showed a reduction wave with a reduction peak potential, E_p^c, of -1.37 V (*vs.* Al in [emim]Cl-AlCl₃, X(AlCl₃) = 0.667, ionic liquid)[16], with a corresponding oxidation wave appearing when the scan was reversed at -1.60 V following the reduction wave (Figure 2a). The peak potential separation, ΔE_p, for this couple was approximately 0.070 V, which is close to the theoretical value of 0.062 V expected for a one-electron, reversible electrode process at 40 °C. Furthermore, the peak current ratio, i_p^c / i_p^a, was found to be constant and close to unity. A similar U(IV)/U(III) reversible couple has been observed by Anderson *et al*[17] for UCl₄ dissolved in a basic [emim]Cl-AlCl₃ (X(AlCl₃) < 0.50) ionic liquid. Furthermore, visible/near infrared spectroscopic data, as well as potentiometric data recorded by Anderson and co-workers, indicated that both the uranium (IV) and uranium(III) species exist as hexachloroanions.

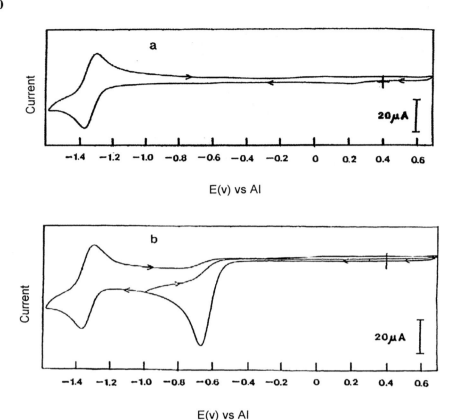

Figure 2. Cyclic voltammograms recorded at a stationary glassy carbon disc electrode at 40 °C in the [emim]Cl-AlCl₃ (X(AlCl₃) = 0.444) ionic liquid:(a) [emim]₂[UCl₆] and (b) [emim]₂[UO₂Cl₄]. The scan rates were 50 mV s⁻¹, and the initial potentials were 0.40 V.

The salt $[emim]_2[UO_2Cl_4]$ was dissolved in the basic ionic liquid to produce a pale yellow solution. A cyclic voltammogram of this solution recorded at a scan rate of 50 mV s⁻¹ showed two voltammetric reduction waves; the first has an E_p^c of -0.66 V and is approximately twice the height of the second reduction wave at E_p^c = -1.37 V. The voltammetric half-wave potential, $E_{1/2}$, for this redox couple, was estimated from E_p^c and the anodic peak potential, E_p^a, $[E_{1/2} = (E_p^c + E_p^a)/2]$ was found to be -1.34 V (Figure 2b). When the scan was reversed at -1.00 V, following the first reduction wave, no oxidation current was observed.

The second reduction wave showed a corresponding oxidation current when the scan was reversed at -1.60 V. In addition, ΔE_p for this redox process was 0.070 V, which is in good agreement with the theoretical value of 0.062 V at 40 °C. In addition, i_p^c / i_p^a was found to be close to unity. No corresponding oxidation wave was observed following the first reduction wave even at scan rates of up to 100 V s^{-1}. Controlled potential electrolysis experiments conducted at an applied potential of -1.00 V showed that the first reduction wave corresponded to a two-electron process. A cyclic voltammogram recorded in this solution after electrolysis was identical to that recorded in a solution containing [emim]$_2$[UCl$_6$]. These results suggest that the two-electron reduction process is accompanied by fast transfer of oxide ion to the IL, with subsequent formation of the hexachlorouranate ion, [UCl$_6$]$^{2-}$, according to eq. (1).[18]

$$[UO_2Cl_4]^{2-} + 2[AlCl_4]^- + 2e^- \rightarrow 2"\{AlOCl_2\}^{-"} + [UCl_6]^{2-} + 2Cl^- \qquad (1)$$

Electrochemistry of [emim]$_2$[UBr$_6$] and [emim]$_2$[UO$_2$Br$_4$] in the basic [emim]Br-AlBr$_3$ ($X = 0.444$) ionic liquid at 40 °C.

The salt [emim]$_2$[UBr$_6$] was dissolved in the basic [emim]Br-AlBr$_3$ ionic liquid to produce a light green solution. A cyclic voltammogram of this solution that was recorded at a scan rate of 50 mV s^{-1} showed a reduction wave with $E_p^c = -1.13$ V (*vs.* Al in [emim]Br-AlBr$_3$, X(AlBr$_3$) = 0.667, ionic liquid) with a corresponding oxidation wave appearing when the scan was reversed at -1.50 V following the reduction wave (Figure 3a). $E_{1/2} = -1.10$ V for this redox couple and $\Delta E_p = 0.070$ V, which is close to the theoretical value of 0.062 V expected for a one-electron reversible redox process at 40 °C. Data collected at a rotating glassy carbon disc electrode (RGCDE) confirmed this $E_{1/2}$ value and verified that this was a one-electron redox process.

The salt [emim]$_2$[UO$_2$Br$_4$] was dissolved in the basic [emim]Br-AlBr$_3$ ionic liquid to produce a pale yellow solution. A cyclic voltammogram recorded in this solution at a scan rate of 50 mV s^{-1} showed two reduction waves similar to those seen during cyclic voltammograms recorded in solutions containing the related chloro species. The reduction wave at $E_p^c = -0.62$ V was approximately twice the height of the second reduction wave at $E_p^c = -1.13$ V (Figure 3b). When the scan was reversed at -0.90 V following the first reduction wave, no oxidation current was observed, even at scan rates of up to 200 V s^{-1}. The second reduction wave showed an associated oxidation wave when the scan was reversed at -1.50 V. The $E_{1/2}$ of this redox couple was -1.09 V. In addition, ΔE_p was 0.065 V for this redox process, which is in excellent agreement with the theoretical value of 0.062 V at 40 °C for a one-electron, reversible electrode reaction.

In order to determine the number of electrons associated with the first reduction wave and the products resulting from this reduction process,

52

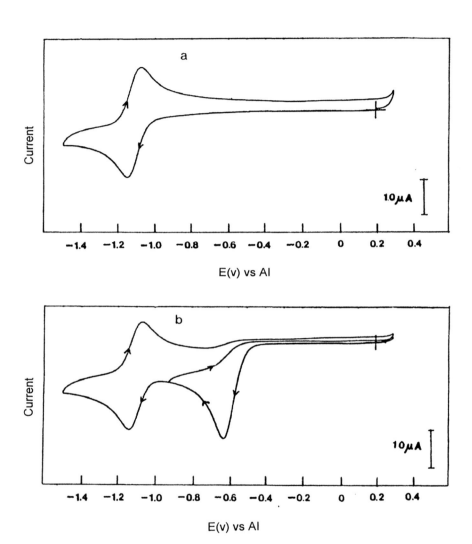

Figure 3. Cyclic voltammograms recorded at a stationary glassy carbon disc electrode at 40 ^{O}C in the [emim]Br-AlBr$_3$ (X(AlBr$_3$) = 0.444) ionic liquid: (a) [emim]$_2$[UBr$_6$] and (b) [emim]$_2$[UO$_2$Br$_4$]. The scan rates were 50 mV s^{-1}, and the initial potentials were 0.20 V.

controlled-potential electrolysis experiments were conducted with a solution of $[UO_2Br_4]^{2-}$ of known concentration (8×10^{-5} mol dm^{-3}) in a glassy carbon crucible at an applied potential of -0.80 V. Unfortunately, some residual electrolysis of the ionic liquid also occurred at this potential. As result of this complication, an indirect method of data analysis had to be employed. This method is based on the assumption that during the initial phase of the controlled-potential electrolysis experiment the main contribution to the total reduction current is the current for the reduction of $[UO_2Br_4]^{2-}$. Thus, extrapolation of the initial linear portion of a plot of $\log(i_t)$ versus t,[19] can be used to estimate the charge consumed by the reduction of $[UO_2Br_4]^{2-}$. By using this method, the number of electrons consumed during the reduction of this species was found to be 2.4 ± 0.2. Thus, like the related chloro-species, $[UO_2Br_4]^{2-}$ undergoes a two-electron reduction process. Furthermore, a cyclic voltammogram of the solution after electrolysis was identical to a voltammogram recorded in a solution prepared by dissolving $[emim]_2[UBr_6]$. These results suggest that, like the chloro-derivative, this two-electron reduction process involves the fast transfer of oxide ion to the ionic liquid, with the subsequent formation of the hexabromouranate ion, $[UBr_6]^{2-}$, according to eq. (2).

$$[UO_2Br_4]^{2-} + 2[AlBr_4]^- + 2e^- \rightarrow 2''\{AlOBr_2\}^{-''} + [UBr_6]^{2-} + 2Br^- \qquad (2)$$

Electrochemistry of $[emim]_2[UBr_6]$ and $[emim]_2[UO_2Br_4]$ in the acidic $[emim]Br$-$AlBr_3$ ($X(AlBr_3) = 0.667$) ionic liquid at 40 °C.

The salt $[emim]_2[UBr_6]$ was dissolved in the acidic ionic liquid to produce a dark green solution. A cyclic voltammogram recorded at a stationary glassy carbon disc electrode at a scan rate of 50 mV s^{-1} showed a reduction wave with $E_p^c = 0.84$ V (*vs.* Al in $[emim]Br$-$AlBr_3$, $X(AlBr_3) = 0.667$, ionic liquid) and a corresponding oxidation wave at $E_p^a = 0.96$ V after the scan was reversed at 0 V (Figure 4a). This corresponds to $E_{1/2} = 0.90$ V for the redox process. ΔE_p for this couple is 0.120 V, which is approximately twice the theoretical value at 40 °C. Increasing the scan rate caused ΔE_p to increase, suggesting that this electrode reaction is quasireversible. However, the voltammetric current function, $i_p^c/v^{1/2}$, and i_p^c/i_p^a were constant over the range of scan rates used (0.05 to 200 mV s^{-1}), indicating that the U(IV)/U(III) couple was not complicated by coupled homogeneous chemical reactions.

The limiting currents, i_l, determined from the voltammetric waves obtained at a RGCDE at 40 °C in this same solution (0.0103 mol dm^{-3}) over the range of rotation rates, ω, from 104 to 261 rads^{-1}, increased linearly with $\omega^{1/2}$ as predicted by the Levich equation for a mass transport-controlled reaction (Table I). The diffusion coefficient for U(IV) was calculated from this data and found to be $2.9 \times 10^{-7}\,\mathrm{cm^2\,s^{-1}}$.

The salt $[\mathrm{emim}]_2[\mathrm{UO_2Br_4}]$ was dissolved in the acidic ionic liquid to give initially a dark brown solution, which quickly turned deep green. The cyclic voltammogram of this green solution was identical to that obtained by dissolving $[\mathrm{emim}]_2[\mathrm{UBr_6}]$ in this same ionic liquid (Figure 4b).

Table I. RGCDE limiting current data for $[\mathrm{emim}]_2[\mathrm{UBr_6}]$ and $[\mathrm{emim}]_2[\mathrm{UO_2Br_4}]$ in the $[\mathrm{emim}]\mathrm{Br}\text{-}\mathrm{AlBr_3}$ ($X(\mathrm{AlBr_3})$ = 0.667) ionic liquid as a function of the square root of the rotation rate, ω, at 40 °C.

$[\mathrm{emim}]_2[\mathrm{UBr_6}]^a$		
$i_1 / \mathrm{A} \times 10^{-6}$	$\omega^{1/2} / \mathrm{rad}^{1/2}\,\mathrm{s}^{-1/2}$	$i_1 / \omega^{1/2} / \mathrm{A}\,\mathrm{s}^{1/2}\mathrm{rad}^{-1/2} \times 10^{-6}$
31.0	10.23	3.0
35.0	11.44	3.0
38.0	12.53	3.0
40.5	13.53	3.0
43.0	14.47	3.0
48.0	16.18	3.0
$[\mathrm{emim}]_2[\mathrm{UO_2Br_4}]^b$		
$i_1 / \mathrm{A} \times 10^{-6}$	$\omega^{1/2} / \mathrm{rad}^{1/2}\mathrm{s}^{-1/2}$	$i_1 / \omega^{1/2} / \mathrm{A}\,\mathrm{s}^{1/2}\mathrm{rad}^{-1/2} \times 10^{-6}$
30.5	10.23	2.9
33.5	11.44	2.9
36.5	12.53	2.9
39.5	13.53	2.9
41.5	14.47	2.9
46.5	16.18	2.9

[a] concentration = 0.0103 mol dm^{-3}; [b] concentration = 0.0109 mol dm^{-3}

This result indicates that the dissolution of $[\mathrm{emim}]_2[\mathrm{UBr_6}]$ or $[\mathrm{emim}]_2[\mathrm{UO_2Br_4}]$ in the acidic $[\mathrm{emim}]\mathrm{Br}\text{-}\mathrm{AlBr_3}$ ($X(\mathrm{AlBr_3})$ = 0.667) ionic liquid ultimately produces the same electroactive species. Potentiometric measurements have previously shown the existence of $[\mathrm{AlCl_4}]^-$ ions in acidic ($X(\mathrm{AlX_3})$ > 0.5) haloaluminate melts. Furthermore, EXAFS[20] and mass spectrometry[21] studies have shown that the highly Lewis acidic species, $[\mathrm{AlCl_4}]^-$, present in the acidic haloaluminate melts, can displace halide ions from transition metal centres, resulting in the formation of species of the type $[\mathrm{M(AlCl_4)_3}]^{x-}$. A similar situation can thus be imagined in the analogous acidic bromo-analogue, in which the electroactive species present in this study upon solvation is actually $[U(AlBr_4)_3]^+$. This electroactive species then undergoes a slow charge transfer (quasireversible) reaction at a stationary glassy carbon disc electrode at negative potentials, eq. (3).

$$[U(AlBr_4)_3]^+ + e^- \rightleftharpoons [U(AlBr_4)_3] \tag{3}$$

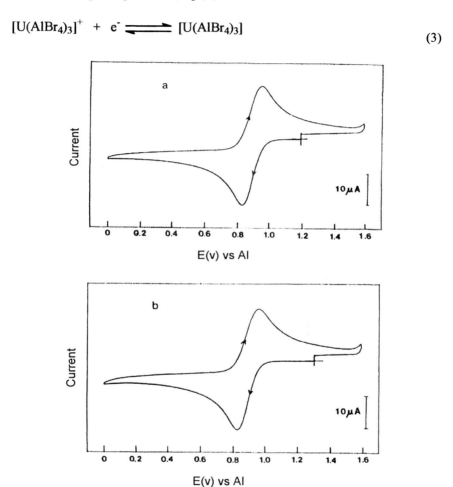

Figure 4. Cyclic voltammograms recorded at a stationary glassy carbon disc electrode at 40 °C in the acidic [emim]Br-AlBr₃ (X(AlBr₃) = 0.667) ionic liquid: (a) [emim]₂[UBr₆] and(b) [emim]₂[UO₂Br₄]. The scan rates were 50 mV s⁻¹, and the initial potentials were 1.20 and 1.30 V, respectively.

Crystallographic Analysis and Discussion

Previous work from our group has shown that use of the van der Waals cutoff criterion to identify C-H$\cdots$Cl hydrogen bonds, overlook many such interactions. The charge transfer nature of hydrogen bonding implies that the directing force of hydrogen contacts is still significant at distances larger than classically accepted – hydrogen bonding is still significant beyond the van der Waals cutoff distance. Due to the directing nature of charge transfer, it is reasonable to assume that a true hydrogen bond occurs in a linear rather than perpendicular orientation. Furthermore to find possible hydrogen bonds in crystal structures, an angular cutoff $\alpha_{C-H\cdots X} > 90°$ is a satisfactory method to seperate hydrogen bonds from non-bonded interactions, since this angular arrangement means that the hydrogen donor atom does, at least, not point away from the acceptor atom. Owing to the directionality of hydrogen bonds, we have previously proposed the replacement of the van der Waals cutoff criterion for determining the presence of hydrogen bonds by a distance/angle based analysis.

Very recently, we expanded upon the previous work in our group (investigating the existence of the C-H$\cdots$Cl bond) by developing a statistical, isotropic density corrected (IDC) method to analyse a variety of contacts. This method involves removing random, non-interacting isotropic density from empirical distribution data, obtained from contacts available from the Cambridge Structural Database (CSD),[22] to expose the non-isotropic behaviour of contacts analysed. This method can be thought of as a "background correction", which removes the isotropic non-bonded atomic distribution at distances far away from the hydrogen atom and thus normalises both the angular and distant accumulation of random contacts. By performing this correction on an experimental contact distribution of data obtained from the CSD, the non-isotropic density, or region within which "real" hydrogen contacts are most likely to occur, can be observed. Subsequently, three dimensional histograms, in the form of IDC contour maps (IDC CMs), representing the relationship between angle, distance and excess density of the contacts can be constructed to visualise the results. For the present discussion of such IDC CMs, a grey scale is used to represent the determined excess densities, *viz.* a region in which no experimental contacts were found is indicated by the colour white, while grey through black represents increasing contact density. In the orginal work, rainbow contour maps were used to display the contact densities, but in the current investigation some of the details are unfortunately lost, since the authors are limited in this black and white publication. Nevertheless, in the present study, contour lines were overlaid on the greyness scale of the IDC CMs, to better emphasise details. By superimposing the distance and angle parameters of a potential hydrogen contact (obtained from a solid state structure) on the

IDC CM, and if these parameters fall within a white region, it is not considered a real hydrogen bond. On the other hand, if the same superimposed distance/angle data falls within a dark grey to black area on the IDC CM, the contact possibly has a directing effect on the crystal structure and is therefore considered to have a high probability of being a hydrogen bond. In the current investigation, this contact data superimposition approach was employed to identify probable hydrogen bonds in the crystalline states of both [emim]$_2$[UCl$_6$] and [emim]$_2$[UO$_2$Cl$_4$].

Crystal structure and hydrogen bonding interactions in [emim]$_2$[UCl$_6$].

The salt [emim]$_2$[UCl$_6$] crystallised in the orthorhombic Pbca space group with the aggregate consisting of two 1-ethyl-3-methylimidazolium cations and one hexachlorouranate(IV) anion (Figure 5).

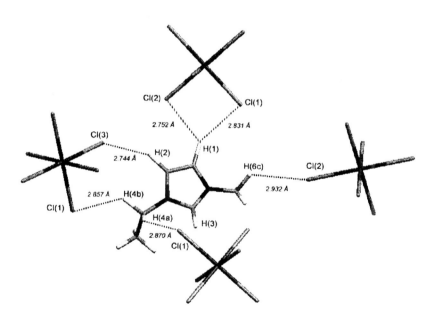

Figure 5 C-H⋯ClU hydrogen bonding interactions for [emim]$_2$[UCl$_6$] in the solid state.

In the anion, the uranium atom is surrounded by six chlorine atoms in an almost perfect octahedral environment with three crystallographically independent U-Cl distances of 2.610(3), 2.608(4) and 2.616(4) Å, which are in good agreement with the mean (2.623(3) Å) of previously reported values,[23] as well as the mean value of 2.668 Å obtained by us from the CSD.[24] In the cation, the imidazolium ring is essentially planar with no atom deviating from the least-squares plane, through C(1), C(2), C(3), N(1) and N(2), by more than 0.009 Å.

In the solid state structure of [emim]₂[UCl₆], six contacts were identified (Figure 5 and Table II) that qualify as hydrogen bonds according to the classical definition *i.e.* their contact distances lie below the van der Waals cutoff distance (2.96 Å for $r^w_H + r^w_{Cl}$). Furthermore, all six the contacts have contact angles > 121.5°, suggesting that they also have a directing effect in the crystal structure.

Table II. Geometry of C-H···ClU and C-H···O=U bonds of [emim]₂[UCl₆] and [emim]₂[UO₂Cl₄] in the crystalline state.

Salt	No.	C-H···Cl	∠ C-H···Cl / °	C-H···Cl / Å
[UCl₆]	1	C(1)-H(1)···Cl(1)	136.34	2.831
	2	C(1)-H(1)···Cl(2)	126.30	2.752
	3	C(2)-H(2)···Cl(3)	132.83	2.744
	4	C(4)-H(4a)···Cl(1)	146.19	2.870
	5	C(4)-H(4b)···Cl(1)	121.50	2.857
	6	C(6)-H(6c)···Cl(2)	126.23	2.932
Salt		C-H···Cl	∠ C-H···Cl / °	C-H···Cl / Å
[UO₂Cl₄]	7	C(4a)-H(4a)···Cl(2)	164.20	2.787
	8	C(4b)-H(4b)···Cl(2)	142.08	2.891
	9	C(3)-H(3)···Cl(2)	145.36	2.950
	10	C(2)-H(2)···Cl(2)	135.42	2.992
	11	C(6c)-H(6c)···Cl(1)	136.49	3.024
		C-H···O	∠ C-H···O / °	C-H···O / Å
	12	C(1)-H(1)···O	168.33	2.535
	13	C(4)-H(4a)···O	115.52	3.054
	14	C(4)-H(4b)···O	124.15	2.835
	15	C(5)-H(5b)···O	111.57	3.040
	16	C(6)-H(6a)···O	146.64	3.052

The contacts, H(1)···Cl(1), H(1)···Cl(2) and H(2)···Cl(3) (Table II, entries 1-3) have respective contact distances of 2.831, 2.752 and 2.744 Å and contact angles of 136.34, 126.30 and 132.83°. The H(1)···Cl(1) and H(1)···Cl(2) interactions can, therefore, be considered a bifurcated hydrogen bond (Figure 5). The contacts, H(4a)···Cl(1), H(4b)···Cl(1) and H(6c)···Cl(2), have respective contact distances of 2.870 Å, 2.857 Å and 2.932 Å, qualifying as classical hydrogen bonds and also possess C-H$\cdots$Cl-U^{2-} contact angles of 146.19°, 121.50° and 126.23° (Table II, entries 4-6), showing their angular quality.

In order to analyse the angular effect of the six hydrogen contacts in the solid state structure of [emim]$_2$[UCl$_6$], an IDC CM was constructed for C-H$\cdots$Cl-M^{n}Z$_x$ (M = any transition metal; Z = any ligand, x = L/no. of ligands) contacts (Figure 6), and the contact distance/angle parameters of the contacts superimposed on the IDC CM. The white line shown in Figure 6 represents the calculated van der Waals cutoff distance for the C-H$\cdots$ClU interactions (2.96 Å) and the white circles represent the six hydrogen contacts in the solid state of [emim]$_2$[UCl$_6$] (Table II entries 1-6; Figure 6). Figure 6 verifies that the six potential hydrogen bonds identified in the solid state of [emim]$_2$[UCl$_6$], qualify as hydrogen bonds according to our proposed distance/angle analysis as well as to the classical definition.

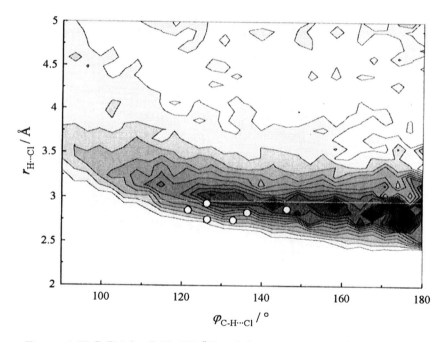

Figure 6. IDC CM for C-H$\cdots$ClM^{n}Z$_x$ solid state contacts in [emim]$_2$[UCl$_6$] showing superimposed distance and angle parameters (o) for H(1)$\cdots$Cl(1), H(1)$\cdots$Cl(2), H(2)$\cdots$Cl(3), H(4a)$\cdots$Cl(1), H(4b)$\cdots$Cl(1) and H(6c)$\cdots$Cl(2). The white line represents the van der Waals cutoff distance.

It is interesting to note that the acidic C(3)-H(3) group is not involved in hydrogen bonding, although the much less acidic C(1)-H(1), C(2)-H(2), C(6c)-H(6c), C(4a)-H(4a) and C(4b)-H(4b) groups are all involved in C-H···ClU hydrogen bonds. Although the contact distances of these interactions are shorter than the van der Waals cutoff distance, using classical hydrogen bonding identification methods, these contacts might have been overlooked since the low acidity of the methylene group decreases the expectation for its involvement in hydrogen bonding, despite the contact lengths falling slightly below the van der Waals cutoff distance.

Crystal structure and hydrogen bonding interactions in [emim]$_2$[UO$_2$Cl$_4$].

The salt [emim]$_2$[UO$_2$Cl$_4$] crystallised in the P2$_{1/n}$ space group with the aggregate consisting of an [UO$_2$Cl$_4$]$^{2-}$ anion and two 1-ethyl-3-methylimidazolium cations (Figure 7). The U-O distance, 1.760(5) Å, and U-Cl distances of 2.661(2) Å and 2.667(2) Å, are in good agreement with other reported values for the *trans*-[UO$_2$Cl$_4$]$^{2-}$ anion, which has been well characterised containing different cations.[25] Furthermore, the U-Cl and U-O bond lengths are in good agreement with the respective mean values of 2.668 and 1.819 Å obtained by us using the CSD. The *trans*-[UO$_2$Cl$_4$]$^{2-}$ anion is in a tetragonally-distorted, six coordinate D$_{4h}$ environment, with four chlorine atoms in the equatorial positions and the two oxygen atoms at the apices. The cation structure displayed nothing unusual with the imidazolium ring being essentially planar with no atom deviating by more than 0.005 Å from the least squares plane through C(1), C(2), C(3), N(1) and N(2).

In the solid state, [emim]$_2$[UO$_2$Cl$_4$] has two types of possible hydrogen bonding interactions, *viz.* C-H···ClU and C-H···O=U. Five potential C-H···ClU as well as five potential C-H···O=U contacts were identified from the single crystal structural data of [emim]$_2$[UO$_2$Cl$_4$] (Table II, Figures 7 and 9). The potential hydrogen bonds were selected since they all exhibited contact angles > 111° and furthermore, all the contact distances are close to the respective C-H···Cl and C-H···O van der Waals cutoff distances of 2.96 and 2.70 Å. Representations of the two possible contact types are shown in Figure 7 (C-H···ClM^{n}Z$_x$) and Figure 9 (C-H···O=M^{n}Z$_x$, (M = any transition metal; Z = any ligand, x = L/no. of ligands), while the IDC CMs of the same two contact types are respectively displayed in Figures 8 and 10. As before, the white lines displayed on the IDC CMs represent the van der Waals cutoff distances for the C-H···Cl (2.96 Å) and C-H···O (2.70 Å) intermolecular hydrogen contacts.

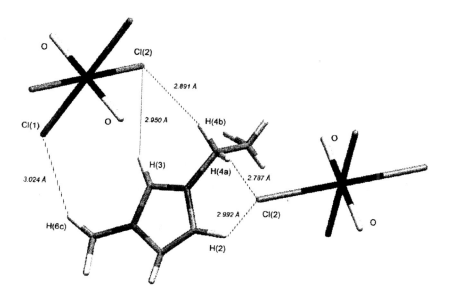

Figure 7. C-H···ClU potential hydrogen bonding interactions for [emim]₂[UO₂Cl₄].

Of the five possible C-H···ClU contacts, the H(4a)···Cl(2), H(4b)···Cl(2) and H(3)···Cl(2) distances of 2.787, 2.891 and 2.950 Å fall below the van der Waals cutoff distance (Table II, entries 7-9), while the respective H(2)···Cl(2) and H(6c)···Cl(1) contact distances of 2.992 and 3.024 Å (Table II, entries 10 and 11) are longer than the der Waals cutoff length. Therefore, classically only the H(4a)···Cl(2) and H(4b)···Cl(2) contacts would have been identified as hydrogen bonds. Similarly, according to classical hydrogen bonding identification methods, the contact lengths of H(2)···Cl(2) and H(6c)···Cl(1) would have excluded their classification as hydrogen bonds. Although only two of the possible five C-H···ClU²⁻ contacts qualify as classical hydrogen bonds, we were interested in establishing the hydrogen bonding character of all five contacts using our statistical analysis method. This was achieved by superimposing the contact distance/angle data of the contacts on the IDC CM (Figure 8) for C-H···ClMⁿ⁻Zₓ contacts.

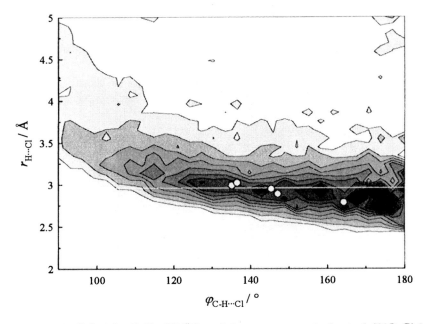

*Figure 8. IDC CM for C-H⋯ClMⁿZₓ solid state contacts in [emim]₂[UO₂Cl₄]
showing superimposed distance and angle parameters (o) for
C(4a)-H(4a)⋯Cl(2), C(4b)-H(4b)⋯Cl(2), C(3)-H(3)⋯Cl(2), C(2)-H(2)⋯Cl(2)
and C(6c)-H(6c)⋯Cl(1). The horizontal white line represents the van der Waals
cutoff distance.*

From Figure 8 it is apparent that all five contacts fall within areas of significant excess density showing the directing effect expected for hydrogen bonds; thus it is proposed that they all can be considered as real hydrogen bonds. Figure 8 further shows that although all five contacts are true hydrogen bonds, the H(3)⋯Cl(2), H(6c)⋯Cl(1) and H(2)⋯Cl(2) and H(6c)⋯Cl(1) contacts are certainly weaker interactions than H(4a)⋯Cl(2) and H(4b)⋯Cl(2) since their contact parameters fall within areas of lower excess density. As opposed to the solid state of [emim]₂[UCl₆] where C(3)-H(3) is not involved in C-H⋯ClU hydrogen bonding, in the crystalline state of [emim]₂[UO₂Cl₄], this acidic hydrogen atom forms a hydrogen bond with a Cl(2) atom. It is again interesting to note that the low acidity H(4a), H(4b) and H(6c) donor atoms are all involved in forming C-H⋯ClU hydrogen bonds.

As already mentioned, in the solid state structure of [emim]$_2$[UO$_2$Cl$_4$], five potential C-H···O=U hydrogen bonds were identified (Figure 9; Table II) by searching for contacts close to the van der Waals cutoff distance (2.70 Å) and possessing the required angular criterion (> 90°).

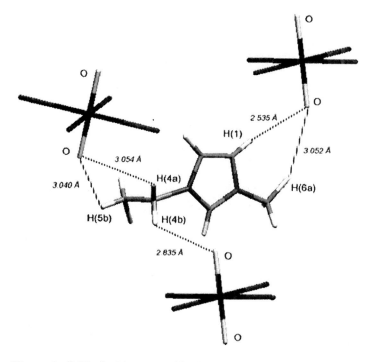

Figure 9. C-H···O=U potential hydrogen bonding interactions for [emim]$_2$[UO$_2$Cl$_4$].

Of these five potential C-H···O=U hydrogen bonds, only H(1)···O is shorter than the van der Waals cutoff distance, qualifying it as a classical hydrogen bond, while the remaining four interactions are too long to be considered classical hydrogen bonds. When the solid state structure of [emim]$_2$[UO$_2$Cl$_4$] was originally determined in our group,[7,26] it was believed that the H(4b)···O interaction with a contact distance of 2.835 Å could effectively be ignored as a hydrogen bond (Figure 9) since the contact distance was too long and in addition, the low acidity of the alkyl proton (compared to the acidic H(2) proton) would not allow it to participate in hydrogen bonding, its position merely being imposed by steric effects. On the other hand, the H(1)···O interaction displayed all the necessary criteria for a hydrogen bond; virtual linearity (168.33°), an acceptor atom containing a lone pair of electrons, and an interaction distance smaller than the sum of the Van der Waals radii of the constituent atoms (2.535 Å). A further complication that arose at the time of the previous study, was that there was no reliable method available to confidently assign hydrogen bonding character to close contacts. Currently, however, by constructing the IDC CM for C-H···O=M^{n}Z$_x$ contacts (Figure 10), and superimposing the distance/angle contact data for the five possible contacts, a more definitive answer as to whether the five potential hydrogen interactions are true hydrogen bonds, can be obtained.

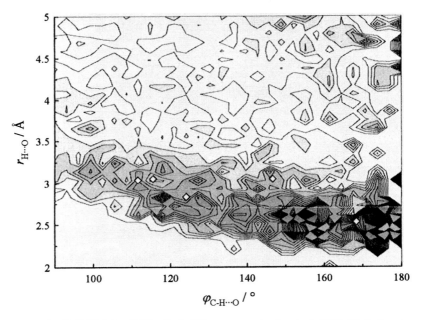

Figure 10. IDC CM for C-H···O=M^{n}Z$_x$ contacts in [emim]$_2$[UO$_2$Cl$_4$] showing superimposed distance and angle parameters (◊) for C(1)-H(1)···O, C(4)-H(4a)···O, C(4)-H(4b)···O, C(5)-H(5b)···O and C(6)-H(6a)···O.

Compared to Figure 8, Figure 10 has a more diffuse character. This is because significantly less C-H···O=M^{n}Z$_x$ contacts are available for analysis in the CSD and thus it appears more dispersed. This presents the immediate problem of confidently assigning C-H···O=M^{n}Z$_x$ contacts for the solid state structure of [emim]$_2$[UO$_2$Cl$_4$]. Of the five potential C-H···O=M^{n}Z$_x$ contacts *viz.* C(1)-H(1)···O, C(4)-H(4a)···O, C(4)-H(4b)···O, C(5)-H(5b)···O and

C(6)-H(6a)···O, the first contact in the list can be confidently identified as hydrogen bonded according to its superimposed distance/angle position on the IDC CM. The remaining four contacts lie too far above the van der Waals cutoff distance, but more importantly, within areas of insufficient excess density to be named as hydrogen bonds.

Summary

This paper has built on a previous communication describing the structures and electrochemistry of 1-ethyl-3-methylimidazolium uranium(IV) and uranium(VI) halide salts and now, more complete details of both these areas are provided.

The electrochemical behaviour of uranium(IV) and uranium(VI) salts was investigated in both acidic and basic haloaluminate ionic liquids. In basic haloaluminate ionic liquids, the uranium(IV) salts showed simple, one-electron reversible U(IV)/U(III) couples, while under the same conditions, the uranium(VI) salts were reduced to uranium(IV) species by an irreversible two electron process with the simultaneous transfer of oxide to the ionic liquid. Dissolution of the uranium(IV) and uranium(VI) salts in acidic haloaluminate ionic liquids indicated that in both cases, the same electroactive specie is present in solution. This result implied that $[AlX_4]^-$ (X = Cl or Br), present in acidic haloaluminate ionic liquids, displaces the original ligands from the uranium metal centre.

Over the past twenty years, the existence and function of hydrogen bonding in the solid state has been addressed more frequently and with increasing importance. Research has focused on the existence or absence of weak hydrogen bonds, which were classically not accepted as interacting contacts. However, in the past, structural chemists did not have any reliable tools available to them to confidently assess the presence or absence of hydrogen contacts in the crystalline state and thus many hydrogen bonding interactions, especially weak interactions, were overlooked.

Classically, the van der Waals cutoff test has been applied to identify hydrogen bonds, and although this method is both simple and easy to apply, it does not take into account the very real, directional nature of all hydrogen bonding interactions. Therefore, we have recently developed a statistical method in our group to search for real hydrogen bonds. In the current investigation, this method was employed for the first time to identify hydrogen bonds in the solid state. This approach distinguishes real hydrogen bonds from van der Waals interactions by virtue of their directional nature. Although this method is only available as a qualitative tool, it represents the first application of a technique to reliably assign true hydrogen contacts. In addition, this method

66

has allowed us to revisit earlier structural studies and re-address our previous misunderstandings of the structural assemblies of uranium halide salts in particular.

Acknowledgements

We (MD and JAvdB) would respectively like to thank QUILL and Sasol for the funding of this study.

Notes and References

1. (a) Fannin, A. A.; King, L. A.; Levisky, J. A.; Wilkes, J. S. *J. Phys. Chem.* **1984**, *88*, 2609-2614; (b) Fannin, A. A.;. Floreani, D. A.; King, L. A.; Landers, J. S.; Piersma, B. J.; Stech, D. J.; Vaughn, R. L.; Wilkes, J. S.; Williams, J. L. *J. Phys. Chem.* **1984**, *88*, 2614-2621.
2. van den Berg, J-A.; Seddon, K. R. *Crystal Growth and Design* **2003**, *3*, 643-661.
3. Holbrey, J. D.; Reichert, W. M.; Nieuwenhuyzen, M.; Johnson, S.; Seddon, K. R.; Rogers R. D., *Chem. Commun.* **2003**, 1636-1638.
4. (a) Holbrey, J. D.; Reichert, W. M.; Swatloski, R. P.; Broker, G. A.; Pitner, W. R.; Seddon, K. R.; Rogers, R. D. *Green Chem.* **2002**, *5*, 407-414; (b) Holbrey, J. D.; Reichert, W. M.; Niewenhuyzen, M.; Sheppard, O.; Hardacre, C.; Seddon, K. R. *Chem. Commun.* **2003**, 476-478; (c) Holbrey, J. D.; Seddon., K. R. *J. Chem. Soc. Dalton Trans.* **1999**, 2133-2141.
5. Abdul-Sada, A. K.; Greenway, A.;M.; Hitchcock, P. B.; Mohammed, T.J. Seddon, K.,R.; Zora, J. A. *J. Chem. Soc. Chem. Commun.* **1986**, 1753-1754.
6. Bondi, A. *J. Chem. Phys.* **1964**, *68*, 441-451.
7. Aakeröy, C. B.; Evans, T. A.; Seddon, K. R.; Pálinkó, I. *New. J. Chem.*,**1999**, *23*, 145-153.
8. Aakeröy, C. B.; Seddon, K.R.; *Chem. Soc. Rev.* **1993**, *22*, 397-407.
9. Umeyama, H.; Morokuma, K. *J. Am. Chem. Soc.* **1976**, *99*, 1316-1341.
10. Desiraju, G. R. *Acc. Chem. Res.* **1991**, *24*, 290-296.
11. Steiner, T. *Chem. Commun.* **1997**, 727-735; Steiner, T. *Angew. Chem., Int. Ed.*, **2002**, *41*, 48-76.
12. Hitchcock, P. B.; Mohammed, T. J.; Seddon, K. R.; Zora, J. A.; Hussey, C. L.; Ward, E. H. *Inorg. Chim. Acta* **1986**, *113*, L25-L26.
13. The composition of a tetrahaloaluminate ionic liquid is best described by the apparent mole fraction {$X(AlX_3)$} of AlX_3 present, where X = Cl or Br. Ionic liquids with $X(AlX_3) < 0.5$ contain an excess of Cl⁻ ions over [Al_2X_7]⁻

ions, and are called 'basic'; those with $X(AlX_3)$ 0.5 contain an excess of $[Al_2X_7]^-$ ions over Cl^-, and are called 'acidic'; melts with $X(AlX_3) = 0.5$ are called 'neutral'.

14. Gale, R. J.; Osteryoung, R. A. In *Molten Salt Techniques;* Lovering, D. G.; Gale, R. J.; Eds.; Plenum, New York, **1983**, Vol. 1, pp 55-78.

15. Seddon, K. R. In *Molten Salt Forum: Proceedings of 5th International Conference on Molten Salt Chemistry and Technology*, Ed. H. Wendt, **1998,** Vol. 4-6, pp. 53-62.

16. The reference electrode of choice for use in both acidic and basic melts is the Al(III)/Al couple in acidic melt, since Al is unstable in contact with basic chloroaluminate melts. This reference electrode consists of a piece of clean Al wire in $X(AlX_3)$ 0.60 or 0.667 melt, which is contained in a small tube terminated with a fritted glass membrane. The tube is inserted into the bulk melt and serves as the reference electrode.

17. Anderson, C. J.; Deakin, M. R.; Choppin, G. R.; D'Olieslanger, W. D.; Heerman, L.; Pruett, D. J. *Inorg. Chem.* **1991**, *30*, 4013-4016.

18. "{AlOCl₂}⁻" and "{AlOBr₂}⁻" refer to a class of hydrolysed aluminium species that are believed to exist in haloaluminate ionic liquids, but which are poorly characterised in solution.

19. Bard, A. J.; Faulkner, L. R. *Electrochemical Methods: Fundamentals and Applications*; Wiley, New York, **1980**, pp. 288-379.

20. Dent, A. J.; Seddon, K. R.; Welton, T. *J. Chem. Soc. Chem. Commun.* **1990**, 315-318.

21. Abdul-Sada, A. K.; Greenway, A. M.; Seddon, K. R.; Welton, T. *Org. Mass. Spectrom.* **1992**, *27*, 648-649.

22. (a) Allen, F. H. *Acta Crystallogr. Sect. B* **2002**, *58*, 380-388; (b) Allen, F. H.; Motherwell, W.D.S. *Acta Crystallogr. Sect. B* **2002**, *58*, 407-422.

23. (a) Bombieri G.; Bangall, K. W. *J. Chem. Soc. Chem. Commun.* **1975**, 188-190; (b) Caira, M. R.; de Wet, J. F.; du Preez, J. G. H.; Gellatly, B. J.; *Acta Crystallogr.* **1978**, *B34*, 1116-1120.

24. The mean value of the bond length was obtained by searching the CSD for all U-Cl bonds in single crystal x-ray structures displaying no disorder, with R < 0.05 and no disoder in the crystal as defined in the software available with the CSD. The distances obtained in this way were then averaged to obtain the mean bond length value.

25. (a) Bois, C.; Dao, N. Q.; Rodier, N. *Acta. Crystallogr.* **1976**, *B32*, 1541-1544; (b) Bombieri, G.; Forsellini, E.; Graziani, R. *Acta. Crystallogr*, **1978**, *B34*, 2622-2624; (c) Moody, D. C.; Ryan, R. R. *Cryst. Struct. Comm.* **1979**, *8*, 933-935; (d) Bombieri, G.; Benetello, F.; Klahne, E.; Fischer, R.D. *J. Chem. Soc. Dalton Trans.* **1983**, 1115-1118.

26. Zora, J.A. *D.Phil.* Thesis, University of Sussex, Sussex, UK, 1986.

Chapter 5

Raman and X-ray Studies on the Structure of [bmim]X (X=Cl, Br, I, [BF$_4$], [PF$_6$]): Rotational Isomerism of the [bmim]$^+$ Cation

Hiro-o Hamaguchi, Satyen Saha, Ryosuke Ozawa, and Satoshi Hayashi

Department of Chemistry, School of Science, The University of Tokyo, Tokyo 113–0033, Japan

Raman spectroscopic and x-ray diffraction studies of [bmim]X (X=Cl, Br, I, [BF$_4$], [PF$_6$]) have revealed that rotational isomers of the [bmim]$^+$ cation exist in the crystalline and liquid states. In monoclinic [bmim]Cl Crystal (1), the *n*-butyl group of [bmim]$^+$ takes a *trans-trans* conformation with respect to the C$_7$-C$_8$ and C$_8$-C$_9$ bonds, while in orthorhombic [bmim]Cl Crystal (2) and [bmim]Br, it takes a *gauche-trans* confor-mation. In liquids, at least two rotational isomers exist, one having a *trans* conformation and the other having a *gauche* conformation around the C$_7$-C$_8$ bond. Co-existence of rotation-al isomers seems to be crucial in hindering crystallization and hence lowering the melting points of [bmim]X crystals.

The 1-butyl-3-methylimidazolium cation, [bmim]$^+$ (Figure 1), makes a number of ionic liquids (ILs) with varying properties, when combined with different anions (1). [bmim]Cl and [bmim]Br are crystals at room temperature, while [bmim]I is a room-temperature ionic liquid (RIL). By cooling down molten [bmim]Cl and [bmim]Br below the melting points, we can easily prepare their super-cooled liquids. Those halogen salts thus comprise a unique system for studying the structure of the [bmim]$^+$ cation in the crystalline and liquid states at room temperature. X-ray diffraction can determine the structures in crystals, while Raman spectroscopy facilitates comparative studies of the structures in crystals and liquids. The two salts, [bmim][BF$_4$] and [bmim][PF$_6$], are prototype RILs that are most extensively used in basic IL investigations as well as in practical applications. Therefore, structure determination of the [bmim]$^+$ cation in these two RILs will be an important first step for the elucidation of the microscopic liquid structure of ionic liquids in general.

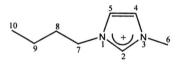

Figure 1. The 1-butyl-3-methylimidazolium cation.

Crystal Polymorphism of [bmim]Cl

We found crystal polymorphism of [bmim]Cl by chance (2). Two different types of crystals, Crystal (1) and Crystal (2), formed when liquid [bmim]Cl was cooled down to –18 °C and was kept for 48 hours (Figure 2). Crystal (2) dominantly formed but Crystal (1) also formed occasionally. Upon leaving Crystal (2) for more than 24 hours at dry-ice temperature, Crystal (2) was converted to Crystal (1). Therefore, it is most likely that Crystal (2) is a metastable form and Crystal (1) is the stable form at this temperature. It is not clear whether Crystal (1) can form directly from the liquid state. Soon after our paper (2) was published, Holbrey et al. reported the same crystal polymorphism (3). They obtained two crystal polymorphs I and II, which correspond to Crystal (2) and Crystal (1), respectively, of our paper.

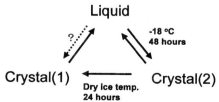

Figure 2. Formation of the two crystal polymorphs of [bmim]Cl.

The powder x-ray diffraction patterns of the two polymorphs of [bmim]Cl, Crystal (1) and Crystal (2), and [bmim]Br are shown in Figure 3. The patterns for [bmim]Cl Crystal (2) and [bmim]Br somewhat resemble each other, while that of [bmim]Cl Crystal (1) is distinct from the other two.

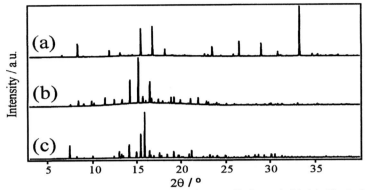

Figure 3. X-ray powder patterns for [bmim]Cl Crystal (1) (a), [bmim]Cl Crystal (2) (b) and [bmim]Br (c).

The Raman spectra of those three salts are compared in Figure 4. Again, the Raman spectra of [bmim]Cl Crystal (2) and [bmim]Br agree very well with each other, while that of [bmim]Cl Crystal (1) is different from the other two. Since the halogen anions are totally inactive in Raman scattering, theses Raman spectra indicate that the [bmim] cation takes the same structure in [bmim]Cl Crystal (2) and [bmim]Br but that it takes a different structure in [bmim]Cl Crystal (1).

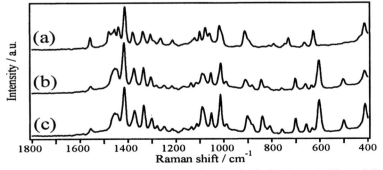

Figure 4. Raman spectra of [bmim]Cl Crystal (1) (a) (upper), [bmim]Cl Crystal (2) (b) and [bmim]Br(c).

Crystal Structures of [bmim]Cl and [bmim]Br

Subsequent to the discovery of the crystal polymorphism of [bmim]Cl, we determined the crystal structures of [bmim]Cl Crystal (1) and [bmim]Br at room temperature (*4,5*). Independently, Holbrey et al. reported the crystal structures of [bmim]Cl Crystal (1) and Crystal (2), and [bmim]Br at −100 °C (*3*). The two sets of structures determined independently at different temperatures agree with each other except for the lattice constants that vary with temperature. They also show that, as already indicated by the Raman spectra in Figure 4, the structure of the [bmim]$^+$ cation in [bmim]Cl Crystal (2) is different from that in (1) but that it is the same as that in [bmim]Br.

The molecular arrangements in [bmim]Cl Crystal (1) and [bmim]Br at room temperature are shown in Figures 5 and 6, respectively. [bmim]Cl Crystal (1) belongs to the monoclinic space group $P2_1/n$ with a=9.982(10), b=11.590(12), c=10.077(11) Å, β=121.80(2)°, while [bmim]Br to the orthorhombic space group $Pna2_1$ with a=10.0149(14), b=12.0047(15), c=8.5319(11) Å. In both structures, the [bmim]$^+$ cations and the halogen anions form separate columns extending along the a axis and no ion pairs exist. The imidazolium rings are all in planar pentagon forms. In [bmim]Cl Crystal (1), the n-butyl group takes a *trans-trans* (*TT*) conformation with respect to the C_7-C_8 and C_8-C_9 bonds, while that in [bmim]Br takes a *gauche-trans* (*GT*) conformation (see also Figure 7). In [bmim]Cl Crystal (1), a couple of the [bmim]$^+$ cations form a pair through a hydrophobic interaction of the stretched n-butyl group. Those pairs stack together and form a column in which all the imidazolium ring planes are parallel with one another. Two types of cation columns with different orientations exist. The imidazolium-ring planes in the two differently oriented columns make an angle of 69.5°. Two straight chains of the anion Cl⁻ directed in the a direction are accommodated in a channel formed by four cation columns, of which two opposite columns have the same orientation. The two chains of Cl⁻ form a zigzag structure as shown in Figure 5-b and the shortest distance between two Cl⁻ in the zigzag is 4.8 Å. This distance is much larger than the sum of the van der Waals radii of Cl⁻ (3.5 Å). There seems to be no specific interaction between two adjacent Cl⁻ anions. In [bmim]Br, only one kind of cation column exists. The imidazolium rings stack so that the N-C-N moiety of one ring interacts with the C=C portion of the adjacent ring. The orientation of the adjacent ring is obtained by rotating a ring plane by about 73° around an axis involving the two N atoms. As in the case of [bmim]Cl Crystal (1), two straight chains of Br⁻, directed in the a direction, are accommodated in a channel produced by four cation columns, which are also extending in the a direction. The shortest Br⁻-Br⁻ distance is 6.55 Å (Figure 6-b) and is again longer than the sum of the van der Waals radii (3.7 Å).

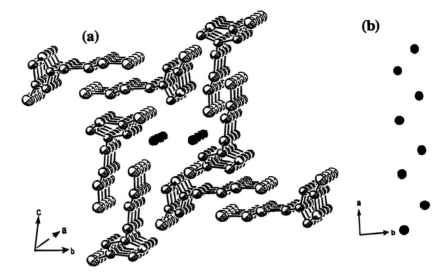

Figure 5. A view of molecular arrangements of [bmim]Cl Crystal (1) in which four cation columns are extending along the a axis (a). Two straight Cl⁻ chains accommodated in the channel formed by the four cation columns (b).

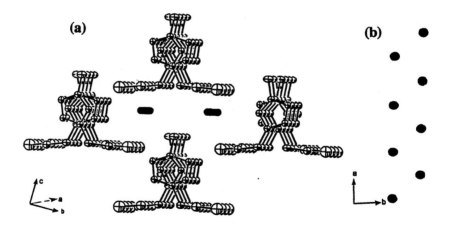

Figure 6. A view of molecular arrangements of [bmim]Br in which four cation columns are extending along the a axis (a). Two straight Br⁻ chains accommodated in the channel formed by the four cation columns (b).

Raman Spectra and Rotational Isomerism of the [bmim]⁺ Cation in [bmim]Cl Crystal (1) and [bmim]Br

The rotational isomerism of the [bmim]⁺ cation is further evidenced by Raman spectroscopy. Figure 7 compares the Raman spectra of [bmim]Cl Crystal (1) and [bmim]Br in the wavenumber region of 400-1000 cm⁻¹. The structural variation of [bmim]⁺ in the two salts, the *TT* and *GT* forms, are also depicted in the same Figure. The two Raman spectra are markedly different from each other. In the wavenumber region 600-700 cm⁻¹, where ring deformation bands are expected, two bands appear at 730 cm⁻¹ and 625 cm⁻¹ in [bmim]Cl Crystal (1) but not in [bmim]Br, while another couple of bands appear at 701 cm⁻¹ and 603 cm⁻¹ in [bmim]Br and not in [bmim]Cl crystal (1). As is shown in the following, these ring deformation bands reflect the conformational variation of the *n*-butyl group. In the 800-900 cm⁻¹ region, the CH₃ and CH₂ deformation bands of the *n*-butyl group are located. The Raman spectral differences in this wavenumber region is also indicative of the rotational isomerism.

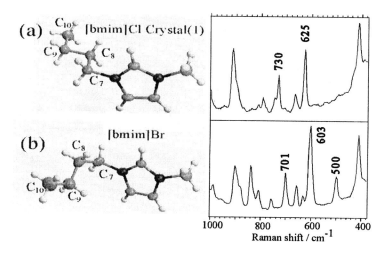

Figure 7. Raman spectra (right) and structure (left) of the [bmim]+ cation in [bmim]Cl Crystal (1) (a) and [bmim]Br (b).

In order to clarify the origins of the observed Raman bands in Figure 7 and their correlation with the rotational isomerism, we carried out a DFT (density functional) calculation with Gaussian 98(6), B3LYP/6-31G+** level. The structures of the *TT* and *GT* forms of the isolated [bmim]⁺ cation were optimized in the vicinity of the structures determined, respectively, for [bmim]Cl Crystal (1) and [bmim]Br by the x-ray analysis. The optimized TT form is found to have lower energy than the GT form by 0.19 kcal/mol. Figure 8 compares the calculated frequencies and intensities (vertical bars) with the observed spectra. The calculation reproduces the observed spectra very well, particularly in the wavenumber region 500-700 cm^{-1}. It shows that the 625 cm^{-1} band of [bmim]Cl Crystal (1) and the 603 cm^{-1} band of [bmim]Br originate from the same ring deformation vibration but that they have different magnitudes of couplings with the CH$_2$ rocking vibration of the C$_7$ carbon. The coupling occurs more effectively for the gauche conformation around the C$_7$-C$_8$ bond, resulting in a lower frequency in the *GT* form (603 cm^{-1}) than in the *TT* form (625 cm^{-1}). Note that the coupling with the CH$_2$ rocking mode having a higher frequency pushes down the frequency of the ring deformation vibration. The same coupling scheme holds for another ring deformation mode and the *GT* form has a lower frequency (701 cm^{-1}) than the *TT* form (730 cm^{-1}). It is therefore elucidated by the DFT calculation that the 625 and 730 cm^{-1} bands are characteristic of the *trans* conformation around the C$_7$-C$_8$ bond, while the 603 and 701 cm^{-1} band are characteristic of the *gauche* conformation. In other words, we can use these bands as key bands to probe the conformation around the C$_7$-C$_8$ bond of the [bmim]$^+$ cation. The 500 cm^{-1} band of [bmim]Br is ascribed to the C$_7$-C$_8$-C$_9$ deformation vibration of the gauche conformation around the C$_7$-C$_8$ bond.

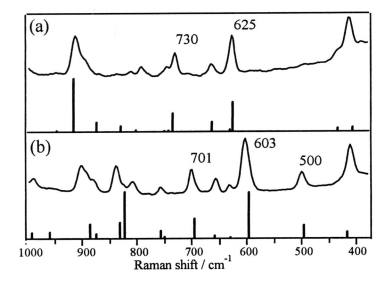

Figure 8. The observed (line) and calculated (vertical bars) Raman spectra of [bmim]Cl Crystal (1)(a) and [bmim]Br(b).

Raman Spectra and Structure of [bmim]⁺ Cation in Liquids

Raman spectra of liquid [bmim]X (X=Cl, Br, I, [BF₄], [PF₆]) are compared in Figure 9. Those of [bmim]Cl Crystal (1) and [bmim]Br are also shown. All Raman spectra were measured at room temperature. The Raman spectra of liquid [bmim]Cl and [bmim]Br were obtained from their supercooled states.

The Raman spectra of the $[BF_4]^-$ and $[PF_6]^-$ anions are already well known. Except for these anion bands which are deleted in Figure 9, the Raman spectra of liquid [bmim]X are surprisingly alike with one another, strongly suggesting that the structure of the [bmim]⁺ cation is very similar in these liquids. Both of the two sets of key bands, 625 and 730 cm⁻¹ bands for the *trans* conformation and the 603 and 701 cm⁻¹ band for the *gauche* conformation, appear in all of the liquid spectra. Therefore, at least two rotational isomers, one having a *trans* conformation and the other having a *gauche* conformation around the C_7-C_8 bond, co-exist in liquid [bmim]X. The relative intensity of the 625 cm⁻¹ band to the 603 cm⁻¹ is a direct measure of the *trans/gauche* ratio. According to the Raman spectra in Figure 9, this *trans/gauche* ratio is very similar for [bmim]Cl, [bmim]Br, [bmim][BF₄] and [bmim][PF₆] but is significantly higher for [bmim]I. Recently, polymorphism in [1-dodecyl-2,3-dimethylimidazolium]PF₆ has been suggested based on DSC measurements (7). It is highly likely that rotational isomers also exist for the [1-dodecyl-2,3-dimethylimidazolium] cation.

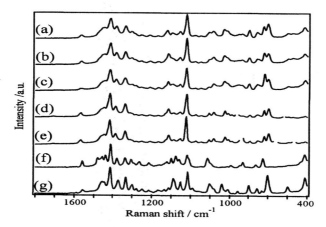

Figure 9. Raman spectra of liquid [bmim]X, where X=Cl(a) , Br (b), I (c), [BF₄] (d), [PF₆] (e), and those of [bmim]Cl Crystal (1) (f) and [bmim]Br(g).

Upon rapidly heating a small piece (0.5 mm x 0.5 mm x 0.5 mm) of [bmim]Cl crystal (1) from room temperature to 72 °C, a droplet of liquid in a non-equilibrium state was transiently formed. It then took more than 10 minutes for this liquid to reach the equilibrium state at 72 °C. The time-resolved Raman spectra in Figure 10 show this melting and equilibration processes. Just after melting, the 625 cm^{-1} band of the *trans* form is much stronger than the *gauche* band at 603 cm^{-1} and the intensity ratio of the two bands becomes constant only after 10 minutes. The unusually long equilibration time strongly suggests that the *trans* and *gauche* conformers of [bmim]$^+$ can be transformed from each other only through a collective motion (analogous to phase transition) of *n*-butyl groups. If the [bmim]$^+$ cation could undergo *trans/gauche* transformation at the single molecular level, the equilibration should occur much faster. Considering the column structures of the [bmim]$^+$ cation found in [bmim]X crystals, we suspect that local structures similar to those exist also in liquid [bmim]Cl.

If we assume such local structures existing also in the other [bmim]X liquids, we can account for the striking similarity of the Raman spectra of those liquids as already shown in Figure 9. It seems that the same molecular structures of [bmim]$^+$, most probably the structures similar to those in the cation columns in crystals, are commonly involved in these ionic liquids. Our most recent large-angle x-ray scattering experiment on liquid [bmim]I shows clear peaks in the radial distribution curve, also indicating the existence of such local structures in the liquid (8). Possibility of local structures in ionic liquids has been pointed out by Optical Kerr-effect spectroscopy (9, 10), small angle X-ray (11), sum frequency generation (12), neutron diffraction (13) and also by theoretical calculations (14, 15, 16). Taking into account the fact that [bmim]X ionic liquids are all transparent, we expect that the dimension of the possible local structure is much smaller than the wavelength of visible light (<100 Å).

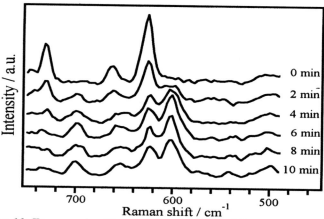

Figure 10. Time-resolved Raman spectra of liquid [bmim]Cl upon melting [bmim]Cl Crystal (1).

Conclusions

Raman spectroscopic and x-ray diffraction studies of [bmim]Cl Crystal (1) and Crystal (2), and crystalline [bmim]Br have revealed that two rotational isomers, the *TT* and *GT* forms of the n-butyl group of the [bmim]$^+$ cation, exist in these crystals. Raman spectra and DFT calculation show that at least two rotational isomers, one having a *trans* and the other having a *gauche* conformation around the C_7-C_8 bond of the *n*-butyl group, co-exist in [bmim]X ionic liquids, where X=Cl, Br, I, [BF$_4$], [PF$_6$]. Time-resolved Raman study of the melting and equilibration processes of [bmim]Cl Crystal (1) suggests that the [bmim]$^+$ cations in those liquids form mesoscopic local structures that are similar to the column structures found in the crystals. Co-existence of two different mesoscopic local structures is likely to hinder crystallization of liquid [bmim]X and hence to lower the melting points of the corresponding crystals. Ionic liquids may not be genuine liquids and they might be better called **nano-structured fluid**. The structural characteristics uncovered in the present study will be of crucial importance in elucidating the many interesting properties of [bmim]X ionic liquids.

References

1. Welton,T. *Chem. Rev.*, **1999**, *99*, 2071.
2. Hayashi, S.; Ozawa, R.; Hamaguchi, H. *Chem. Lett.* **2003**, *32*, 498.
3. Holbrey, J. D.; Reichert, W. M.; Nieuwenhuyzen, M.; Johnston, S.; Seddon. K. R.; Rogers, R. D. *Chem. Commun.* **2003**, 1636.
4. Saha, S.; Hayashi, S.; Kobayashi, A.; Hamaguchi, H. *Chem. Lett.* **2003**, *32*, 740.
5. Ozawa, R.; Hayashi, S.; Saha, S.; Kobayashi, A.; Hamaguchi, H. *Chem. Lett.* **2003**, *32*, 948.
6. Gaussian 98, Rev. A.11.1: Frisch, M. J.; Trucks, G. W.; Schlegel, H. B.; Scuseria, H. B.; Robb, M. A.; Cheeseman, J. R.; Zakrzewski, V. G.; Montgomery, J. A.Jr.; Stratmann, R. E.; Burant, J. C.; Dapprich, S.; Millam, J. M.; Daniels, A. D.; Kudin, K. N.; Strain, M. C.; Farkas, O.; Tomasi, J.; Barone, V.; Cossi, M.; Cammi, R.; Mennucci, B.; Pomelli, C.; Adamo, C.; Clifford, S.; Ochterski, J.; Petersson, G. A.; Ayala, P. Y.; Cui, Q.; Morokuma, K.; Salvador, P.; Dannenberg, J. J.; Malick, D. K.; Rabuck, A. D.; Raghavachari, K.; Foresman, J. B.; Cioslowski,J.; Ortiz, J. V.; Baboul, A. G.; Stefanov, B. B.; Liu, G.; Liashenko, A.; Piskorz, P.; Komaromi, I.; Gomperts, R.; Martin, R. L.; Fox, D. J.; Keith, T.; Al-Laham, M. A.; Peng, C. Y.; Nanayakkara, A.; Challacombe, M.; Gill,P. M. W.; Johnson, B.; Chen, W; Wong, M. W.;

Andres, J. L.; Gonzalez, C.; Head-Gordon, M.; Replogle, E. S.; Pople, J. A.; *Gaussian, Inc.*, Pittsburgh PA, **2001**.

7. Fox, D. M.; Awad, W. H.; Gilman, J. W.; Maupin, P. H.; De Long, H. C.; Trulove, P. C. *Green Chem.*, **2003**, *5*, 724.

8. Katayanagi, H.; Hayashi, S.; Hamaguchi, H.; Nishikawa, K. *Chem. Phys. Lett.* **2004**, *392*, 460.

9. Gerard, G.; Gordon, C. M.; Dunkin, I. R.; Wynne, K. *J. Chem. Phys.*, **2003**, *119*, 464.

10. Hyun, B-R.; Dzyuba, S. V.; Bartsch, R. A.; Quitevis, E. L.; *J. Phys. Chem. A* **2002**, *106*, 7579.

11. Moutiers, B. G.; Labet, A.; Azzi, A. E.; Gaillard, C.; Mariet, C.; Lutzenkirchen, K. *Inorg. Chem.*, **2003**, *42*, 1726.

12. Iimori, T.; Iwahashi, T.; Ishii, H.; Seki, K.; Ouchi, Y.; Ozawa, R.; Hamaguchi, H.; Kim, D. *Chem. Phys. Lett.*, **2004**, *389*, 321.

13. Hardacre, C.; Holbrey, J. D.; McMath, S. E. J.; Brown, D. T.; Soper, A. K. *J. Chem. Phys*, **2003**, *118*, 273.

14. Morrow, T. I.; Maginn, E. J. *J. Phys. Chem. B*, **2002**, *106*, 12807.

15. Shah, J. K.; Brennecke, J. F.; Maginn, E. *Green Chem.* **2002**, *4*, 112.

16. Hanke, C. G.; Price, S. L.; Lynden-Bell, R. M. *Mol. Phys.*, **2001**, *10*, 801.

Chapter 6

In Situ IR Spectroscopy in Ionic Liquids: Toward the Detection of Reactive Intermediates in Transition Metal Catalysis

Ralf Giernoth

Institute of Organic Chemistry, University of Cologne, Greinstrasse 4, 50939 Köln, Germany

In situ IR spectroscopy is used for the *in situ* investigation of the mechanisms of two selected reactions. The *ReactIR* proves to be a versatile tool for reaction monitoring and for the on-line collection of kinetic information. With this method, the direct alkylation of *N*-methylimidazole with 1-bromobutane can be shown to proceed via S_N1 or electron transfer rather than S_N2. For easy and straightforward detection and characterization of catalytically reactive intermediates in the Suzuki-Miyaura reaction, though, the method seems to be not sensitive enough, but it can help to gain insight into the reaction mechanism.

Ionic liquids (ILs) are well established as neoteric reaction media for organic synthesis and for transition metal catalysis in particular (*1*). This is mainly due to the fact that their physical and physico-chemical properties are varied substantially with the choice of anion and cation – vital properties such as miscibility with other organic solvents, solvation power, acidity, gas solubility, or viscosity (*2*). By wisely choosing a variety of ionic liquids with the properties needed for the intended chemical transformation, the chemist, in addition to catalyst, ligands, reaction conditions etc., now has another means for screening and fine-tuning. Therefore, ILs are sometimes called "designer solvents" (*3*).

From the chemical literature of the past few years, it seems obvious that for almost any given chemical transformation a "matching" IL will exist that can be used as a reaction medium and that will lead to a significant enhancement of activity and selectivity. This fact has sometimes been called "ionic liquid effect" (*4*) – primarily because of the lack of rational explanations for this enhancement. Since it it obvious that some kind of IL effect does exist, and since it differs substantially from case to case, it will be impossible to devise a "general IL effect" – there are far too many very different ILs and far too many different reaction types with many different possibilities of solvent-solute interactions. In the beginning of IL chemistry it was believed that the promoting effect of ILs with weakly coordinating anions (such as BF_4^- or $(CF_3SO_2)_2N^-$) is mainly due to the fact that they are polar enough to dissolve all the reactants but do not exhibit a strong solvent shell (*5*). For transition metal catalysis this would lead to "naked" and thus very active catalysts. Later, on the other hand, it could be demonstrated that some ILs can actively participate in the reaction, for example by binding to the transition metal center as a ligand (*6*). And still, there are the cases of reactions that are not promoted by the use of ionic liquids. Therefore, our key questions are: what are the reaction mechanisms of catalytic reactions in ionic liquids? Do they differ to the ones in conventional organic solvents? What are the key intermediates? What promotes the reactions, and what inhibits them? It won't be possible to answer these questions globally, rather will they have to be answered for any single reaction of interest. We therefore believe that it is vital to develop a standard set of readily applicable *in situ* spectroscopic methods for the use in ionic liquid media. To that end, we wondered to what extend *in situ* IR spectroscopy can be applied.

The *ReactIR 4000* spectrometer (manufactured by *Mettler-Toledo*) is an FT-IR spectrometer with an ATR (attenuated total reflection) probe that can be inserted directly into the reaction mixture. With this, it is possible to monitor any given reaction *in situ* under real reaction conditions. The resulting 3D IR plot for the synthesis of [bmim]Br (1-butyl-3-methyl imidazolium tetrafluoroborate) is shown in Figure 1 as an example. Here, *N*-methylimidazole is added to the reaction flask first, then 1-bromobutane, and finally the mixture is heated to 80 °C. At that point, a vigorous and exothermic reaction starts and is finished within minutes. The process can be analyzed afterwards with a software package (*ConcIRT™*) which notes the change in the IR spectra over time. By this, the number of entities is determined that occur in the reaction, together with their corresponding time-concentration-profiles and their calculated IR spectra. Therefore, it is possible to identify reactive intermediates that are not visually detectable in the IR spectra. For our alkylation reaction (cf. Figure 1), Figure 2 shows that an intermediate occurs which can, e.g., be attributed to a carbocation.

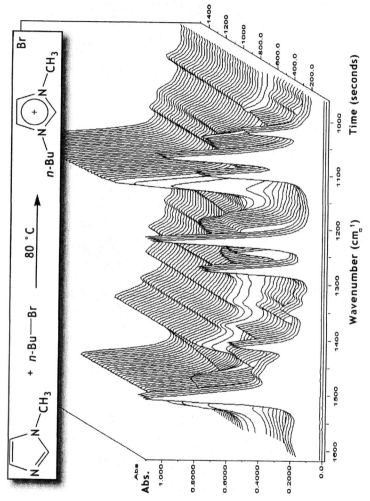

Fiugre 1. ReactIR plot the direct alkylation of N-methylimidazole with 1-bromobutane to give the ionic liquid [bmim] Br.

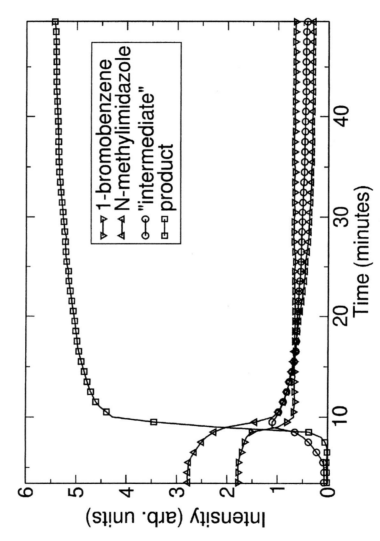

Figure 2. Time-concentration-profile for the reaction from Figure 1, as calculated by the software package ConcIRT.

The reaction seems not to proceed as S_N2 under these reaction conditions, which of course is highly unusual for primary alkyl halides. This can be explained, though, by the fact that 1,3-disubstituted imidazolium-based ILs (one of which is formed as the reaction product) are polar as well as protic (the 2-hydrogen in the imidazolium ring bears some acidity (7)). Polar protic media stabilize charged transition states; therefore, S_N1 or an electron transfer mechanism can be favored here over S_N2.

The *ReactIR* method can conveniently be used for reaction monitoring of transition metal catalysis in ILs. As a test reaction we chose the Suzuki-Miyaura coupling of bromobenzene with phenylboronic acid as published by *Welton et. al. (8)*, which has been reported to be not too sensitive to air. The *ReactIR* monitoring of the standard model reaction in [omim][Tf₂N] (1-octyl-3-methyl imidazolium bis-triflic amine) is shown in Figure 3. The turnover of the reaction is clearly visible, e.g., by the biphenyl bands growing in around 1633 and 1410 cm^{-1} (ν_{C-C}) and around 700 cm^{-1} (δ_{C-H}).

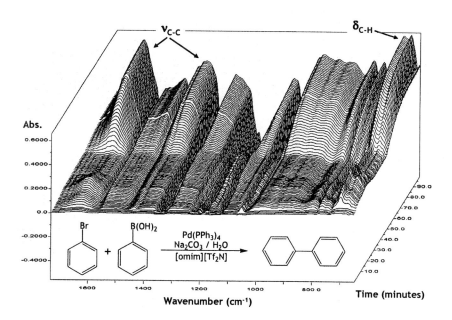

Figure 3. ReactIR monitoring of a standard Suzuki-Miyaura reaction in [omim][Tf₂N]. Three specific bands of the product biphenyl are highlighted.

The *ReactIR* can be used to acquire on-line kinetic information about the reactions. In this fashion, it is possible, e.g., to judge the end of the turnover while the reaction is still running. Figure 4 demonstrates the time-intensity plot for the Suzuki-Miyaura reaction corresponding to Figure 3.

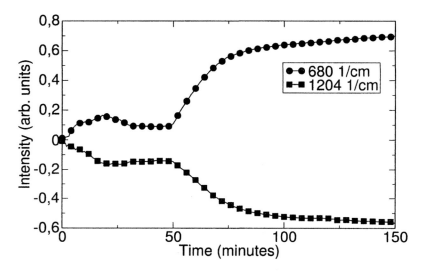

Figure 4. Time-intensity plot for the Suzuki-Miyaura reaction in [omim][Tf₂N] as shown in Figure 3. The intensities of a specific band of one starting material (bromobenzene) and one of the product (biphenyl) in relative values are plotted versus time.

Although being very useful for everyday reaction monitoring, especially on the laboratory scale, *ReactIR* plots of this type do not reveal straightforward information about reactive intermediates in transition metal catalysis. The obvious reason for this is that the spectroscopic method must be fast and sensitive enough for the detection of short-lived intermediates being present at very low concentrations. Compared to NMR, for example, IR spectroscopy is fast: with the *ReactIR* a maximum of three spectra per second can be recorded (although, since it is a FT spectrometer, normally 32 to 64 spectra are accumulated prior to fourier transformation). The sensitivity, on the other hand, is probably not high enough and further limited by the ATR technique.

None the less, we wondered to what extend *in situ* IR spectroscopy of this type can be employed for mechanistic studies of the Suzuki-Miyaura reaction. One of the key questions in this reaction (and in IL chemistry in general) is about the structure of the active catalyst species (cf. Figure 5). Is is well known that imidazolium cations under basic conditions can form carbene complexes of the *Arduengo* type (*7*) with transition metals. *Welton et. al.* have demonstrated that under the reaction conditions a palladium imidazolylidene carbene complex can form, althougth its activity in catalysis could not be proven (*9*). Yet, for a closely related Heck reaction *Xiao et. al.* have shown that a complex of this type is active as a precatalyst (*6*). On the other hand, we realized that tetrakis-tris(triphenylphosphine)palladium(0), which is used as the catalyst, is insoluable in our ILs, but it dissolves readily and completely after the addition of bromobenzene. This is in accordance with the textbook catalytic cycle for this reaction, in which the first step is the insertion of the palladium catalyst into the arene-bromine-bond (Figure 5).

To find out about this and to test the applicability of the *ReactIR* method for this purpose, the following experiment has been carried out: First, a larger amount (50 mg) of the palladium precatalyst has been suspended in the IL (2 mL; in this case, $[C_{10}mim][BF_4]$ has been used, which had proven to be the best solvent for this reaction in our earlier experiments (*10*)). The corresponding *ReactIR* plot (Figure 6) shows at least some dissolved species around 700 cm^{-1}. Then, bromobenzene was added in two portions (100 + 50 µL). At this point, the spectra of bromobenzene are dominant in the IR and nothing else is visually detectable. It is noteworthy to say that under these conditions an almost clear, dark-orange solution was obtained. Keeping the solution under argon overnight resulted in a bright-white triphenylphosphine precipitate.

Subsequent *ConcIRT* analysis gave five (calculated) IR spectra and the corresponding time-intensity plots for five (calculated) reaction partners, as depicted in Figure 7. Number 1 is bromobenzene, since the calculated spectrum corresponds perfectly with the real one, and the intensity plot shows the two steps of addition. No. 2 is the IL solvent – the calculated spectrum is of limited value, because the IL as the solvent has been subtracted from all the other spectra as the background by the *ReactIR*. The intensity plot shows the same, but inverse steps for the addition of bromobenzene, which is a simple dilution effect.

The interpretation of the other three spectra/plots must remain guesswork. No. 3 remains almost constant or diminishes slightly in the course of the reaction while No. 4 and 5 are steadily growing in. Therefore, 3 will most probably have to be

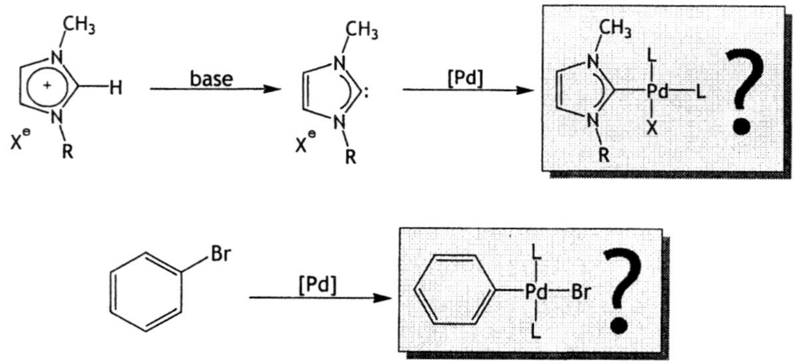

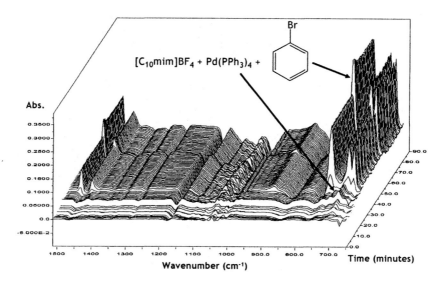

Figure 5. Two possible forms of the active catalyst for Suzuki-Miyaura coupling reactions in an imidazolium-based IL medium.

Figure 6. ReactIR monitoring of the catalyst preforming mechanism.

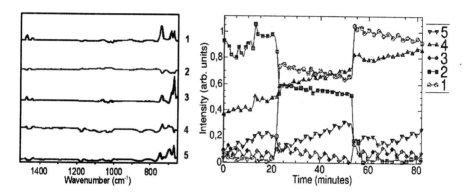

Figure 7. ConcIRT analysis of the ReactIR plot shown in Figure 6. The numbers of the calculated spectra (left) correspond to the numbers of the time-intensity-plots (right). The program calculates five reactants, two of which (No. 1 and 2) can unambiguously be attributed to bromobenzene and the IL, respectively. (The wave-like structure of the time-intensity plots reflects oscillations in the quality of the air generator which is used to flush the spectrometer.)

attributed to Pd(PPh$_3$)$_4$ while 4 and 5 can be PPh$_3$ and the active catalyst component, although it is impossible to say if this is true and which one is which. The fact that dissolution of the Pd-precatalyst only occurs after addition of bromobenzene hints to the fact that under these conditions no carbene complex is formed. Unfortunately, the Pd-carbene stretching band, unambigiously identifying this species, would occur below 650 cm^{-1}, which is the physical limit of the *ReactIR* spectrometer.

In conclusion, we have shown that *in situ* IR spectroscopy with the *ReactIR* is a valuable tool for reaction monitoring and on-line kinetics of reactions in ionic liquids. Reactive intermediates in higher concentration, although invisible by the eye, can be detected by subsequent analysis of the *ReactIR* plot with the software package *ConcIRT*. For easy and straightforward detection of reactive intermediates in transition metal catalysis, though, this method seems to be not sensitive enough. None the less, since the experimental setup is hassle-free and robust, the method can at least help getting ideas about the mechanistic details of the reaction, which will then have to be proven further by other spectroscopic techniques. The author is currently developing a setup for routine *in situ* NMR spectroscopy in ionic liquids.

R.G. would like to thank *Deutsche Forschungsgemeinschaft* (DFG) for a generous *Emmy Noether* fellowship.

References

1. *Ionic Liquids in Synthesis;* Wasserscheid, P.; Welton, T., Eds.; Wiley-VCH: Weinheim 2003; Chapter 5.
2. *Ionic Liquids in Synthesis;* Wasserscheid, P.; Welton, T., Eds.; Wiley-VCH: Weinheim 2003; Chapter 3.
3. Wasserscheid, P.; Keim, W. *Angew. Chem., Int. Ed. Engl.* **2000**, *39*, 3772-3789
4. For example: Ross, J.; Chen, W.; Xu, L.; Xiao, J. *Organometallics* **2001**, *20*, 138-142.
5. Welton, T. *Chem. Rev.* **1999**, *99*, 2071-2083.
6. Xu, L. J., Chen, W. P., Xiao, J. L. *Organometallics* **2000**, *19*, 1123-1127.
7. Arduengo, A. J., Harlow, R. L., Kline, M. *J. Am. Chem. Soc.* **1991**, *113*, 361-363.
8. Mathews, C. J.; Smith, P. J.; Welton, T., *Chem. Commun.* **2000**, 1249-1250.
9. Mathews, C. J.; Smith, P. J.; Welton, T.; White, A. J. P.; Williams, D. J., *Organometallics* **2001**, *20*, 3848-3850.
10. Giernoth, R.; unpublished results.

Chapter 7

In Situ IR Spectroscopic Study of the CO_2-Induced Swelling of Ionic Liquid Media

Nikolaos I. Sakellarios and Sergei G. Kazarian*

Department of Chemical Engineering and Chemical Technology, Imperial College, London SW7 2AZ, United Kingdom

In situ ATR-IR spectroscopy has been used to investigate the effect of high pressure CO_2 on ionic liquid media, since it allows for the simultaneous measurement of two important phenomena (swelling and gas sorption) while it also provides molecular level insight into the behavior of each constituent of the mixture in question. More specifically, in this report, data are presented relating CO_2 pressure with the swelling of room temperature ionic liquids [RTILs] based on the 1-butyl-3-methyl-imidazolium cation combined with the hexafluorophosphate anion. The absorbance of the IR bands of [bmim][PF_6] was used to calculate the ionic liquid swelling and data up to 72 bar at 25°C were recorded and consecutively compared with literature data, while the v_3 band (2335 cm^{-1}) of dissolved CO_2 was used to determine quantitatively its sorption into the ionic liquid medium.

The realization that the ecosystem is deteriorating due to the use and generation of hazardous substances (e.g. volatile organic compounds) by the plethora of the industrial processes currently in operation, has led to the advent of "green chemistry". A comparatively new concept, "green chemistry" represents a fundamental shift in the approach to chemical production and process plant designing. Furthermore, the collaboration between engineers and chemists involved in "green chemistry" aims at devising new, environmentally friendly, synthesis roots for existing chemicals, which utilize novel alternatives to ordinary solvents. According to Professor Seddon at Queen's University Belfast (2) there are four alternatives, with the most promising being supercritical fluids and ionic liquids, without that meaning that solvent-free synthesis or the use of water as a solvent would not be more appropriate in some cases (3).

In the past few years, supercritical fluids (SCFs), have earned significant attention from the scientific community involved in the development of cleaner production processes, due to reduced catalyst contamination as well as improved selectivity and efficiency compared with conventional solvents (4). On the other hand, ionic liquids (ILs) are currently attracting significant attention as novel solvents for the development of new "green" technologies due to their fascinating properties; including the ability to tailor them for specific applications, their excellent chemical and thermal stability (5) and their extended liquid range [approx. 300°C for commercially available ionic liquids] (3, 6). However the most intriguing features of ionic liquids is their non-volatile and highly solvating nature, allowing them to be used in deep vacuum systems without any mass loss and also to readily solvate a wide range of organic, inorganic and organometalic compounds (7, 8). Examples of technologies that could benefit from the unique combination of properties presented by ionic liquids are organic synthesis (9), separations (10) and electrochemistry (11) to mention a few. Nevertheless, prior to making the transition between laboratory and industrial scale applications, it must be ensured that sufficient studies have been conducted to address the effect of ionic liquids on living organisms. The reason for this is that although utilizing ionic liquids ameliorates air pollution, it is almost inevitable that at some stage there could be release of small amounts into environment, including the water supplies. It is also possible that pollution may be involved in the manufacturing of ionic liquids.

However, the full potential of utilizing these two "innovative" alternatives to common solvents was realized only when scientist decided to combine them. By combining ionic liquids and supercritical fluids, one can take advantage of the intriguing properties of both media, hence making the resulting chemical processes even more efficient and environmentally friendly. Recent work has demonstrated that CO_2 is highly soluble in certain ionic liquids (12-16), an observation extending the potential applications of IL in separating CO_2-soluble reaction products from IL by the use of SCF/liquid extraction (*17*). Additionally, recent work has also reported on liquid phase volume expansion of ionic liquids with the introduction of large amounts of CO_2 (*1*). The *in situ* spectroscopic approach developed by our group provides effective means to investigate these phenomena in more detail and hence try to provide insight at the molecular level.

Theoretical Background

In single reflection ATR-IR spectroscopy, IR light is caused to be internally reflected at the interface between a spectroscopic crystal and the sample. Hovever, radiation does in fact penetrate a short distance into the rarer medium, as shown in Figure 1. This penetrating radiation is characterized as an evanescent electric field and its intensity subsides exponentially as it penetrates deeper into the sample, according to the equation below (*18*):

$$E = E_0 \cdot e^{-z/d_p} \tag{1}$$

where E is the electric field amplitude at a penetration distance z into the sample, E_0 is the electric field amplitude at the interface and d_p is the penetration depth given by:

$$d_p = \frac{\lambda}{2\pi \cdot \left(n_1^2 \sin^2 \theta - n_2^2 \right)^{1/2}} \tag{2}$$

where λ is the wavelength of the incident beam, θ is the incident angle and n_1 and n_2 are the refractive indices of the ATR-IR crystal and the sample respectively.

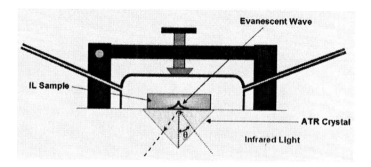

Figure 1: Schematic view of the ionic liquid on the ATR-IR

However since all quantitative calculations are consistently based upon the molar absorptivity used in transmission spectroscopy, in order to apply the Beer-Lambert law [Equation 6] for analyzing the data obtained, a pathlength giving the same absorbance in transmission as that obtained by ATR-IR spectroscopy is required. This pathlength is labeled effective thickness, d_e, and it is dependent on the polarization of the incident light (18, 19). Thus, in the case of p–polarized, the effective thickness is:

$$d_{e.//} = \frac{\lambda(\cos\theta)n_1 n_2 \left[2n_1^2 \sin^2\theta - n_2^2\right]}{\pi\left[n_1^2 \sin^2\theta - n_2^2\right]^{1/2}\left(n_1^2 - n_2^2\right)\left[\left(n_1^2 + n_2^2\right)\sin^2\theta - n_2^2\right]} \tag{3}$$

while if it is s–polarized, the effective thickness is:

$$d_{e,\perp} = \frac{\lambda_1 n_{21} \cos\theta}{\pi\left(1 - n_{21}^2\right)\left(\sin^2\theta - n_{21}^2\right)^{1/2}} \tag{4}$$

Consequently in the case of unpolarized light, the average effective thickness is taken to be as:

$$d_{e.u} = \frac{d_{e.//} + d_{e,\perp}}{2} \tag{5}$$

Finally, the equation describing the Beer-Lambert Law is given as:

$$A = \varepsilon \cdot c \cdot d_{e,u} \tag{6}$$

where A is the absorbance, ε is the molar absorptivity of the sample, c is the concentration of the absorbing species and $d_{e,u}$ is the effective thickness for unpolarised light.

Experimental

Materials

In this work a commercially available room temperature ionic liquid [SACHEM Inc.] based on the 1-butyl-3-methylimidazolium [bmim]$^+$ cation combined with the hexafluorophosphate [PF$_6$]$^-$ anion was used. Since [bmim][PF$_6$] is moderately hydrophilic, prior to use, the IL sample was treated for 8 hours at 80°C and the sample was considered to be moisture free since there was no spectroscopic evidence of the presence of water in the treated sample. Carbon dioxide was supplied by BOC with a nominal purity of 99.9% and was used without further purification.

Apparatuses

In this research, the implementation of the ATR-IR technique in the study of high pressure CO_2/ionic liquid systems has been achieved by using a specially designed cell, which was compatible with the single reflection ATR accessories [Specac Ltd, UK]. More specifically, this method is based on a modified ATR accessory, called the "Golden Gate", which incorporates a diamond internal reflection element, with an incident angle of 45° and ZnSe focusing lenses [Specac Ltd, UK] and a miniaturic high-pressure flow cell as shown in Figure 2. This experimental configuration and methodology has recently received significant interest from the spectroscopic community, as it allows to measure spectra of gases dissolved in various mediums at various pressures (up to 170 bar) and temperatures (the "Golden Gate" plate incorporates a heating element in the base of the ATR crystal, hence allowing for experiments to be carried out at elevated temperatures). All the ATR FT-IR spectra were acquired using an Equinox-55 spectrometer [Bruker Optics, Germany], utilizing a high sensitivity mercury cadmium tellurium (MCT) detector at a resolution of 2 cm^{-1}.

Experimental Principles

Choice of spectral bands to be studied

As shown in Figure 3, the IR bands of pure CO_2 and IL are clearly separated. ATR-IR spectra of IL subjected to CO_2 were recorded for pressures up to 72 bar at room temperature (ca. 25 °C) using a diamond crystal. Figure 3 illustrates the spectral changes that were observed as the pressure of the CO_2 increased. The intensity of all ionic liquid bands decreased while the characteristic absorption bands of the CO_2 sorbed in the ionic liquid medium, v_3 at 2335 cm^{-1} and v_2 at 655 cm^{-1}, increased.

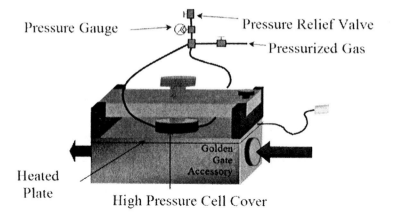

Figure 2: Schematic arrangement of the experimental configuration based on an ATR spectroscopic accessory

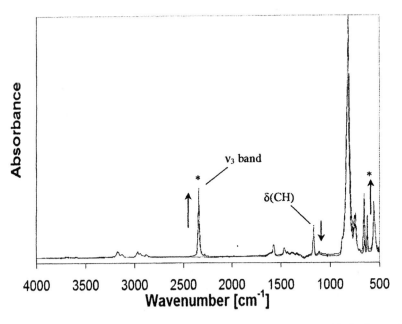

Figure 3: ATR-IR spectra of [bmim][PF6] at ca. 25 °C: before and during exposure to CO2. Absorption bands of CO2 are denoted by an asterisk and the arrows show the trend followed by each band as CO2 pressure increases.

For the analysis in this paper, we will concentrate on the ν_3 band of CO_2 (2336 cm^{-1}) and upon the δ(CH) band of the ionic liquid (1166 cm^{-1}). The reason for selecting these two band is that the ν_3 band of CO_2 (2336 cm^{-1}) is characteristic of CO_2 sorbed in material, while the δ(CH) band of the ionic liquid is an isolated one that does not overlap with other bands, hence making the quantification much easier. These two bands are shown in Figure 4.

As observed from Figure 4, as CO_2 pressure increases the absorbance of the CO_2 bands increases while the absorbance of the ionic liquid band decreases. Measurements are taken at each pressure until equilibrium is achieved. Equilibrium is confirmed by constant absorbance of both CO_2 and ionic liquid bands between two subsequent measurements with a time lag of about 15 minutes. By analyzing the intensity of both CO_2 and IL absorption bands it is possible to determine both the extent of swelling of the IL as well as the gas sorption in the ionic liquid medium.

Calculation of swelling

Assuming that the refractive index and the molar absorptivity, ε , of the 1166 cm^{-1} band do not change with the concentration of the sorbed CO_2, the absorbance of the C-H bending bond of the ionic liquid before (A_o) and after exposure (A) to the gas will be given by:

$$A^o = \varepsilon \cdot c^o \cdot d_{e,u} \qquad (7)$$

$$A = \varepsilon \cdot c \cdot d_{e,u} \qquad (8)$$

where c^o and c are the concentrations of the ionic liquids before and after exposure to gas and $d_{e,u}$ are the effective thickness of unpolarized light. (for the definition of $d_{e,u}$, vide supra section 5.3.3.2).

Another point to mention is that the absorbance of the C-H bonds was measured using the area under the peak and not the peak height in order to avoid any discrepancies due to the shape of the band being affected by pressure.

Assuming that the ionic liquid sample occupies a volume V before exposure and a volume V+ΔV during and after exposure, the following equation defines the swelling S:

$$\frac{c^o}{c} = \frac{V + \Delta V}{V} = 1 + \frac{\Delta V}{V} = 1 + S \qquad (9)$$

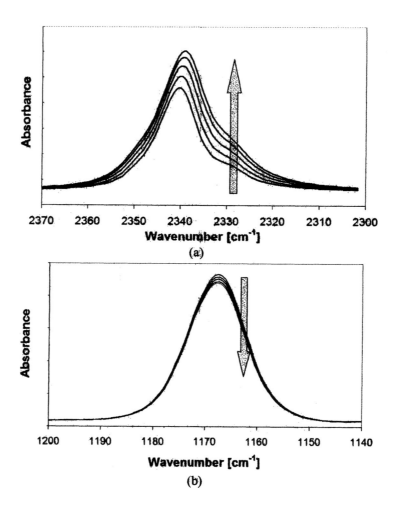

Figure 4: ATR-IR spectra of (a) v_3 band of CO_2 and (b) $\delta(CH)$ band of the ionic liquid. The arrows show the trend followed by each band as CO_2 pressure increases.

which when combined with equations (7) and (8) gives:

$$S = \frac{A^0}{A} * \frac{d_{e,u}}{d^0_{e,u}} - 1 \qquad (10)$$

where S is the swelling A^0 and A the absorption prior and after CO_2 sorption in the IL, and $d_{e,u}$ and $d^0_{e,u}$ are the effective thickness of unpolarized light before and after CO2 sorption. However since we have assumed that the refractive index is unchanged, Equation 10 can be rewritten as:

$$S = \frac{A^0}{A} - 1 \qquad (11)$$

Furthermore, it is possible to apply the Beer-Lambert law to the CO_2 band at 2335 cm^{-1} to calculate the concentrations of CO_2 in the ionic liquid medium using the same molar absorptivity for CO_2 as for CO_2 dissolved in other solvents at high pressures, which has been calculated to be ca. 1.0×10^6 cm^2·mol^{-1} (20).

Results and discussion

Swelling & Sorption:

Using Equation 11 along with the values obtained from the spectra acquired for the CO_2/IL system one can calculate the extent of swelling against the various pressures of CO_2. The plot of the result obtained is provided in Figure 5.

However, since the above results have been acquired based on the assumption that the refractive index of the sample does not change as CO_2 pressure increases, Near-IR transmission spectroscopy has been utilized to validate them because results obtained by this method are independent of changes in the refractive index. Figure 6 shows the spectra obtained by

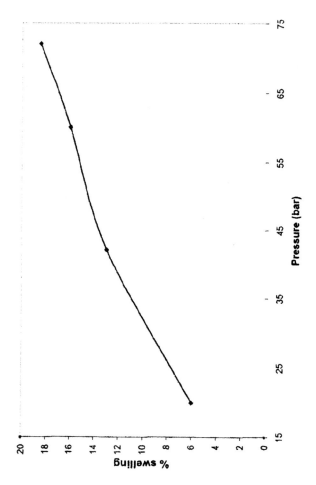

Figure 5: Swelling of [bmim][PF₆] as a function of CO₂ pressure at 25°C

transmission spectroscopy for the ionic liquid bands at 6216 cm^{-1} at 25°C. The results obtained for both the extent of swelling and the amount of CO_2 sorbed into the ionic liquid calculated by our experiment are in good agreement with our data obtained by ATR spectroscopy as well as with literature data (*1, 21*), especially at low pressures. Figure 7 shows how CO_2 sorption data compare to literature values.

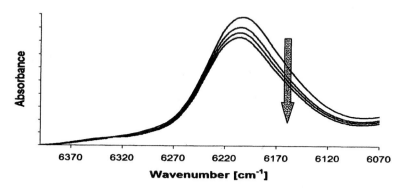

Figure 6: Transmission Near-IR spectra of the ionic liquid under high pressure CO_2 at 25°C. The arrow shows the trend as CO_2 pressure increases.

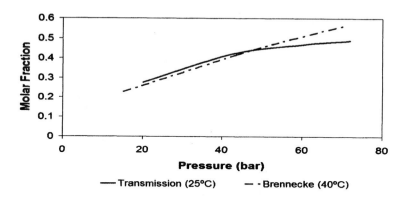

Figure 7: Molar fraction of CO_2 in [bmim][PF$_6$] at 25°C for various pressures

Lu et al.(21) in their recent study, applied UV/vis spectroscopy to monitor the solvatochromatic behavior of the *N,N*-dimehthyl-4-nitroaniline indicator in ionic liquids and high pressure CO_2 mixtures and also measured the extent of volume expansion for these mixtures. Our approach, illustrated in this paper, directly monitors swelling by measuring the absorbance changes of the intrinsic bands of ionic liquids.

Conclusions

In situ spectroscopic approach has been developed which allows the simultaneous measurement of the swelling of an ionic liquid medium and the corresponding gas sorption in the medium, while simultaneously allowing for molecular level insight into the interactions between the ionic liquid medium and high pressure CO_2. The key feature of this approach is the monitoring of absorbance changes of bands directly related to each of the sample constituents by means of a modified ATR-IR accessory, which enables the acquisition of spectra for ionic liquid subjected to high-pressure gas. This method was initially used in the study of PDMS (22), but its application in the field of ionic liquids verified the prediction of the authors that it is a generic method which may be applied to other systems, subject to the constraint that suitable isolated bands for both the sample and the gaseous solute are available.

ATR-IR spectra of [bmim][PF$_6$] ionic liquids subjected to CO_2 were recorded for pressures up to 72 bar at temperatures of 25 °C. Analysis of the absorbance of the v_3 band of CO_2 dissolved in [bmim][PF$_6$] at 2335 cm^{-1} provided quantitative sorption data, while the absorbance of the δ(CH) band of the ionic liquid at 1167 cm^{-1} was used to calculate the ionic liquid swelling. Data obtained by this approach concurs with other reports in the literature(1, 21), indicating that [bmim][PF$_6$] swells approximately by 18% under CO_2 pressure of 72 bar. Currently, supplementary research is underway in our laboratory in order to assess the effect of ionic liquid structure (i.e. length of alkyl chains, different combinations of anions and cations) on the extent of swelling.

Finally, another advantage presented by the outlined technique is the simultaneous measurement of the sorption of various gases in the ionic liquid medium and the resulting swelling of the latter medium; data that can in turn be used in further elucidating the swelling mechanism of ionic liquids. Knowledge that will permit close control of the ionic liquid viscosity, hence opening new horizons in enhancing the rate of reactions utilizing these innovative solvents.

Acknowledgements

We thank *EPSRC* for financial support, *Specac Ltd* and *Bruker Optics* for support and help with the spectroscopic equipment, *SACHEM Inc.* for the provision of the ionic liquid samples.

References

1. Brennecke, J. F.; Blanchard, L. A.,Gu, Z. *Journal of Physical Chemistry B* **2001**, 105, 2437-2444.
2. Seddon, K. R. Ionic Liquids. http://www.nature.com/nsu/000608/000608-15html.
3. Lemetais, C. *The Chemical Engineer* **2000**, 14-15.
4. Rayner, C. M.; Oakes, R. S.,Clifford, A. A. *Journal of the Chemical Society Perkin Transactions 1* **2001**, 917-941.
5. Bonhote, P.; Graetzel, M.; Dias, A. P.; Papageorgiou, N.,Kalyanasundaram, K. *Inorganic Chemistry* **1996**, 35, 1168-1178.
6. Welton, T. *Chemical Reviews* **1999**, 99, 2071.
7. Wasserscheid, P.; Gordon, C. M.; Hilgers, C.; Muldoon, M. J.,Dunkin, I. R. *Chemical Communications* **2001**, 1186–1187.
8. Seddon, K. R. *Journal of Chemical Technology and Biotechnology* **1997**, 68, 351-356.
9. Wasserscheid, P.,Keim, W. *Angewandte Chemie International Edition* **2000**, 39, 3772.
10. Visser, A. E.; Swatloski, R. P.,Rogers, R. D. *Green Chemistry* **2000**, 2, 1.
11. Hussey, C. L., *Chemistry of Non-aqueous Solutions*; Weinheim: 1994; 'Vol.' p.
12. Niehaus, D.; Philips, M.; Michael, A.,Wightman, R. M. *Analytical Chemistry* **1991**, 63, 1728.
13. Niehaus, D.; Philips, M.; Michael, A.,Wightman, R. M. *Journal of Physical Chemistry A* **1989**, 93, 6232.
14. Kazarian, S. G.; Briscoe, B. J.,Welton, T. *Chemical Communications* **2000**, 2047-2048.
15. Kazarian, S. G.; Sakellarios, N. I.,Gordon, C. M. *Chemical Communications* **2002**, 1314-1315.
16. Brennecke, J. F.; Anthony, J. L.,Maginn, E. J. *Journal of Physical Chemistry B* **2001**, 105, 10942-10949.
17. Gordon, C. M. *Applied Catalysis A* **2001**, 222, 101-117.
18. Harrick, N. J., *Internal Reflection Spectroscopy*; Wiley-Interscience: New York, 1976; 'Vol.' p.
19. Urban, M. W., *Attenuated Total Reflectance Spectroscopy of Polymers*; Polymer Surfaces and Interfaces Series American Chemical Society: Washington, DC, 1996; 'Vol.' p.
20. Brill, T. B.; Schoppelrei, J. W.,Maiella, P. G. *Applied Spectroscopy* **1999**, 53, 351-355.
21. Lu, J.; Liotta, C. L.,Eckert, C. A. *Journal of Physical Chemistry A* **2003**.
22. Flichy, N. M. B.; Kazarian, S. G.; Lawrence, C. J.,Briscoe, B. J. *Journal of Physical Chemistry B* **2002**, 106, 755-9.

Chapter 8

Dynamics of Fast Reactions in Ionic Liquids

Alison M. Funston and James F. Wishart

Chemistry Department, Brookhaven National Laboratory,
Upton, NY 11973

Pulse radiolysis and laser flash photolysis are complementary
tools for studying fast reactions in ionic liquids. Both
techniques have been used to study solvation processes in
quaternary ammonium ionic liquids, which extend into the
nanosecond regime and influence the reactivity and energetics
of radiolytically-generated excess electrons. The preparation
and properties of ionic liquid families designed to further these
studies is reported.

Introduction

Ionic liquids form a new class of solvents which have potential uses in
academic as well as industrial settings. One advantage of ionic liquids is that
they may be tailored by selective combination of different cations and anions to
meet the specific needs of an application or technique. A possible application of
ionic liquids may be in the nuclear industry. For example, the incorporation of
boron or chlorine, good thermal neutron poisons, into an ionic liquid could
substantially reduce the risk of nuclear criticality in fuel cycle and radiological
waste handling and decontamination operations (*1*). Experiments on several

imidazolium ionic liquids have found their radiochemical stability to be high (2). In order to determine the primary radiolytic species formed in the liquids and their subsequent reactivity, the technique of pulse radiolysis is ideal (3). In this technique a short pulse of ionizing radiation is focused into the solution of interest, producing radical species. The reactions of these species, either in the presence or absence of added solute, may be followed using a variety of techniques such as transient UV-visible and NIR spectroscopy, conductivity or electron spin resonance.

The dynamics of very fast processes such as solvation and radical or excited state reactions may be resolved using the LEAF pulse radiolysis facility (4, 5). The general processes occurring upon pulse radiolysis of neat ionic liquids have been described in detail previously (6), in addition the primary radiation chemistry (6-8) and the reactivity of the primary species formed in the ionic liquid methyltributylammonium bis((trifluoromethyl)sulfonyl)imide (9-12) has been described in detail. More recently fluorescent solvatochromic probes such as C153 have also been employed to study the solvation effects in ionic liquids. It was found in these and other studies of the dynamics of radical and excited state reactions in ionic liquids (13, 14) that diffusion and solvation processes in the ionic liquids occur on longer timescales when compared to the usual molecular solvents. This is due in some part to the more viscous nature of the ionic liquids, however it has also been found that the reactions occur at a rate faster than predicted by the diffusion of the reactants through the solvent using the modified Debye equation (15, 16), Equation 1, where the ratio of the radii of the reacting species is approximated as 1:1.

$$k_{diff} = 8000RT/3\eta \qquad (1)$$

Despite the recent flurry of activity investigating fast dynamics, to date detailed investigations have focused either on one liquid or one class of ionic liquids, usually imidazolium salts. In an effort to begin to understand the effects of the ionic liquid structure on the initial radiolytic products formed, their solvation dynamics and the diffusion of solutes in ionic liquids we have initiated a program involving the synthesis of series of liquids and their investigation using the LEAF facility in concert with photolysis studies incorporating fluorescent probes.

One current disadvantage of working with ionic liquids is the low fluidity of many of the ionic liquids synthesized to date, which prohibits certain types of investigations (for example experiments in which it is necessary to flow a solution through a cell as is often required in pulse radiolysis), and limits the general usefulness of the liquids. Our interest in performing these types of experiments led to a synthetic program aimed at producing several series of related, low melting point, low viscosity ionic liquids. The characterization of the viscosity and conductivity properties of these ionic liquids enables some understanding of the relationship between the structure and physical characteristics. In addition, the ability to combine the properties of a specific cation with those of an anion allows the design of ionic liquids that generate specific radicals during radiolysis. This may be of use when designing liquids for use in the radiolytic fuel cycle or for radiological decontamination and cleanup, as it would allow the conversion of radiolysis products into specific channels that augment the radiation stability of the liquid.

Designing Ionic Liquids for Radiolysis Studies

Selection of Candidate Liquids on the Basis of Reactivity and Properties

The interaction of high energy radiation with liquids causes excitation and ionization of the liquid molecules, resulting ultimately in the formation of a number of energetic, "excess" electrons and a corresponding number of electron vacancies or "holes" in the solvent. The fate of the electrons is dependent upon the solvent, which may itself be reactive towards the electrons. However if the solvent's reactivity is low the electrons may survive a sufficient amount of time to become localized and then solvated, and they may also react with added solutes. The most frequently studied ionic liquids contain imidazolium or pyridinium cations that have more positive reduction potentials than that of the solvated electron; thus the electron is very rapidly scavenged to form a neutral radical (17). Since the reactivity and spectral and dynamic properties of the excess electron are of primary interest in these studies, quarternary ammonium ionic liquids, in which the solvated electron is stable, have been a focus of the synthetic program. To promote the formation of low melting, low viscosity ionic liquids, the bis((trifluoromethyl)sufonyl)imide anion ([NTf$_2$]) (18-20) was employed.

Pyrrolidinium and quaternary ammonium cations have enjoyed prominence recently in the field of ionic liquids and are known to form low melting ionic liquids, particularly with the [NTf$_2$]$^-$ anion (20, 21). The replacement of one of the alkyl chains within these well-known cations with an ether-containing chain of the same number of atoms leads to a dramatic reduction in the melting point and viscosity of the ionic liquid (see Table 1) as has been observed previously for quaternary ammonium salts (22-24) and alkylsulfonate anions (25). Indeed, replacing the pentyl chain in N-methyl,N-pentyl-pyrrolidinium cation ([C$_5$mpyrr]$^+$) with an ethoxyethyl group results in the bromide salt being liquid even at room temperature. In this instance, incorporation of a halide anion such as Br$^-$ in the liquid would be likely to produce [Br$_2$]$^{-\cdot}$ radicals upon radiolysis, a strongly oxidizing species that can be useful for studying oxidation-induced electron transfer and radical reactions in ionic liquids (13).

Table 1. Selected physical properties of the ionic liquids at 25 °C

Ionic Liquid	m.p., °C	Density, g cm^{-3}	Conductivity, mS cm^{-1}	Walden Product P S cm^2 mol^{-1}
[C$_4$mpyrr]Br	209			
[C$_5$mpyrr]Br	153			
[Rmpyrr]Br	63			
[R'mpyrr]Br	< 15	1.31	0.11	1.7
[C$_4$mpyrr][NTf$_2$]	-18 [a]	1.40	2.7	0.61
[C$_5$mpyrr][NTf$_2$]	< 10	1.36	2.1	0.60
[Rmpyrr][NTf$_2$]	< 10	1.46	3.9	0.61
[R'mpyrr][NTf$_2$]	< 10	1.41	3.4	0.56
[NBu$_4$]Br	102 [b]			
[NBu$_3$R]Br	58			
[NBu$_4$][NTf$_2$]	89 [c]			
[NBu$_3$R][NTf$_2$]	43	1.22 [g]	1.4 [g]	0.43 [g]
[NBu$_3$Hx][NTf$_2$]	26 [d]	1.19	0.16 [d]	0.45
[NHMe$_3$][NTf$_2$]	80 [e]	1.74 [h]		
[NHEt$_3$][NTf$_2$]	3.5 [f]	1.43	5.2	0.66
		1.40 [g]	10.8 [g]	0.56 [g]
[NHPr$_3$][NTf$_2$]	49	1.30 [g]	4.5 [g]	0.41 [g]
[NHBu$_3$][NTf$_2$]	36	1.30 [g]	2.0 [g]	0.26 [g]

R = $CH_2CH_2OCH_3$, R' = $CH_2CH_2OCH_2CH_3$. Data from reference (26) unless otherwise indicated.

[a] MacFarlane et al (20, 27-29).

[b] Aldrich catalog value.

[c] Reported as 91 °C by Matsumoto et al (18).

[d] Sun et al (30).

[e] Forsyth et al. (31).

[f] Susan et al. (32).

[g] At 50 °C.

[h] Crystallographic (solid) density (31).

Another series of liquids of intense interest for their potential applications as fuel cell electrolytes and support media for photocatalytic H_2 generation contain trialkylammonium cations in combination with the [NTf$_2$]$^-$ anion. Upon radiolysis, the cations in these liquids capture excess electrons to produce the hydrogen atom in high yield (26). They provide a cleaner route to the study of H-atom reactivity in ionic liquids than the acidified tetraalkylammonium liquids previously studied (8). In those liquids, H-atoms were only produced in the presence of water, which is known to change the properties of the liquids (33).

Properties

The melting points of the ionic liquids used are presented in Table 1 along with their densities and conductivities at 25 °C. As may be seen from the data, the substitution of an alkyl chain by an ether containing chain of the same number of atoms lowers the melting point of pyrrolidinium ionic liquids by as much as 140 °C. However for the tertiary alkyl ammoniums the difference created by ether substitution is much less, being of the order of 40 °C although much larger reductions of the melting point as a result of ether substitution have been reported previously for other tertiary alkyl ammoniums (22, 23). The conductivity of the pyrrolidinium liquids is also increased by about 1 mS cm^{-1} by the substitution, consistent with earlier observations (22, 23, 34, 35).

The liquids in the proton-donating trialkylammonium series display melting points that are lower then would be expected based upon their tetraalkyl analogues (21). Furthermore, they exhibit quite high conductivities, higher than any reported for aprotic ionic liquids, a phenomenon that has been reported previously for ionic liquids containing monoalkylammonium cations (32, 36, 37).

The conductivities of most of the ionic liquids listed in Table 1 are high by virtue of their low viscosities (see below). Normalized ionic conductivity is expressed in the form of the Walden product (viscosity × molar volume × conductivity). The "ideal" value (34, 36) for the Walden product of a liquid consisting of monovalent ions (based on dilute KCl$_{aq}$) would be 1.0 P S cm^2 mol^{-1}; values for typical ionic liquids (33) vary from 0.4 to 0.9 P S cm^2 mol^{-1}. A classification diagram based on the Walden Product has been used to assess the ionicity of ionic liquids. The value of 1.7 P S cm^2 mol^{-1} for [R'mpyrr]Br (where R' = CH$_2$CH$_2$OCH$_2$CH$_3$) suggests that this liquid may be superionic (34). The series of protonated tertiary amines shows a trend of decreasing Walden products as the alkyl chain length increases, which indicates a decreasing degree of proton transfer in the series (36) and consequently lower ionicities.

Viscosities

Viscosity measurements show that the ionic liquids derived from the fully saturated cations used in these studies cover a very large viscosity range at room temperature. The liquids can be classified into three groups on the basis of their viscosity-temperature profiles shown in Figure 1. The highest viscosity liquids, [R'mpyrr]Br and [NBu$_3$Hx][CH$_3$SO$_3$], are literally as viscous as honey. Diffusion will be very slow in these liquids, so for the purpose of photolytic and radiolytic kinetics studies they can be considered to be room-temperature glassy matrix isolation media. The substitution of methanesulfonate anion ([CH$_3$SO$_3$]$^-$) in place of [NTf$_2$]$^-$ raises the viscosity tremendously (12900 vs. 910 cP at 20 °C, respectively) in agreement with the comparison of [NTf$_2$]$^-$ liquids and their non-fluorinated analogues (38).

The next class of liquids includes tetraalkylammonium salts of [NTf₂]⁻ and dicyanamide ([N(CN)₂]⁻), with viscosities between 500 and 1000 cP at 20 °C. The tetradecyl(trihexyl)phosphonium salts of [NTf₂]⁻ or [N(CN)₂]⁻ also fall into this category, with a slightly weaker viscosity dependence on temperature than the ammonium liquids. These liquids are fluid enough to be conveniently handled for the bulk of the radiolysis studies, including pre-solvated and solvated electron capture (*6, 7*), reactions of hydrogen atoms with aromatics (*8*), radical reactions (*9, 10*) and redox reactions (*11, 13*).

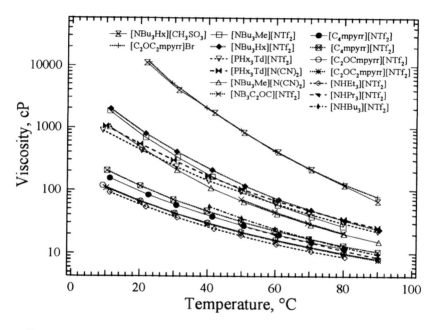

Figure 1. Temperature dependence of the viscosities of several ionic liquids (26). Each liquid was dried in a vacuum oven for 48 hours at 60 °C. The viscometer was continuously purged with dry N₂ during the measurements.

The final class of ionic liquids, ranging from 60 to 120 cP at 20 °C, includes the pyrrolidinium salts (*20*) and their ether derivatives, and the series of [NTf₂]⁻ salts of protonated tertiary ammines (*32*). These liquids are fluid enough to be used in flow apparatus commonly used for fast kinetics studies, such as the picosecond pulse-probe transient absorption system at LEAF (*4, 5*), which requires sample exchange to avoid cumulative radiation effects. In future experiments, these liquids will allow direct spectroscopic observation of the electron solvation process by monitoring the decay of pre-solvated electrons that absorb in the near infrared with picosecond resolution.

Solvation Dynamics in Ionic Liquids

One of the key properties of a solvent when it is used as a reaction medium is the response of the solvent to the movement of charge. Charge movement is a very common occurrence in chemical reaction mechanisms, not only in the obvious cases of redox reactions and charge transport, but also in any reaction where the polarity of the transition state differs from that of the reactants. As ionic liquids are finding increasing applications in photoelectrochemical cells, batteries, fuel cells and as reaction media for a wide range of reactions, it is important to understand the factors that control their solvation dynamics. Standard methods of analyzing solvation dynamics use time-resolved spectroscopy of solvatochromic probes that have been excited into non-equilibrium configurations. Our investigations include transient absorption studies of radiolytically-produced excess electron solvation and emission studies of Stokes shift and polarization anisotropy decay of coumarin-153 measured in the same ionic liquids. The parallel nature of the two experiments is shown in Figure 2.

Following radiolysis, when the electron comes to rest after losing excess kinetic energy from the initial ionization event, it is delocalized and mobile (middle top of Fig. 2). Fluctuations in the ionic bath cause the electron to become weakly trapped, then the trap deepens as the ionic lattice reorganizes around the electron. As the trap deepens the absorption spectrum of the electron species shifts to higher energy. Eventually the electron is localized in the lattice, possibly analogous to the F-center in ionic solids (top right, displaced anion out of frame). The electron solvation is thus coupled to translational diffusion in the ionic liquid through the displacement of ions required for solvation, and thus to the bulk viscosity of the solvent.

Pulse radiolysis: probe by absorption of electron species.

Laser photolysis: probe by fluorescence (Stokes shift, anisotropy decay).

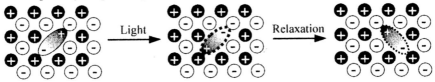

Laser excitation of fluorescent dyes with large dipole moment differences between their ground and excited states, such as coumarin-153, is shown in the lower part of Figure 2. Photoinduced charge redistribution creates an excited dye molecule surrounded by a non-equilibrium distribution of ions (bottom middle). Coulombic repulsion induces solvent ion translation and dye molecule reorientation to lower the energy of the system (bottom right), which is observed as a red shift of the emission spectrum with time. The solvation process couples to ion translational diffusion and bulk viscosity as in the electron case. In addition, it is reasonable to expect the rotational dynamics measured by emission polarization anisotropy to correlate well with the spectrally observed solvation dynamics.

Photolytic Studies of Solvation Dynamics of Ionic Liquids

The dynamic solvation of common solvatochromic probes, such as coumarin-153 (39-42), 6-propionyl-2-dimethylaminonaphthalene (PRODAN) and 4-aminophthalimide (4-AP, Figure 3), have been used recently to assess the response of imidazolium based ionic liquids to changes in the charge distribution of the probes upon excitation.

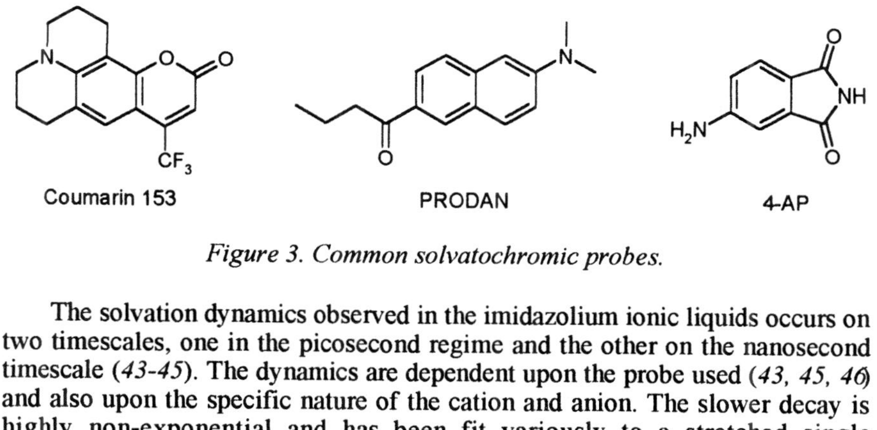

Coumarin 153   PRODAN   4-AP

Figure 3. Common solvatochromic probes.

The solvation dynamics observed in the imidazolium ionic liquids occurs on two timescales, one in the picosecond regime and the other on the nanosecond timescale (43-45). The dynamics are dependent upon the probe used (43, 45, 46) and also upon the specific nature of the cation and anion. The slower decay is highly non-exponential and has been fit variously to a stretched single exponential (43, 44) and to two double exponential decays (46-48), generally this component has been ascribed to the large scale reorganization of the solvent structure, presumably involving coupled reorganization of the cation and anion (43-48) and correlates well with the solvent viscosity (44). The faster decay cannot be resolved using these standard techniques, as the instrument response times are typically of the order of tens of picoseconds. However, the existence of a faster process is indicated by the shift of the earliest observed emission spectrum from the position of the estimated time-zero spectrum (49). To this end, optical Kerr-effect spectroscopy has been employed (*50-52*). The fast time scale response has been attributed to local librational movements of the imidazolium cation (*43, 44, 50*).

In contrast to the imidazolium ionic liquids, very little is known about the dynamics of ammonium based liquids. Early molten salt studies on tetraalkylammonium salts containing perchlorate and hydrogen sulfate as the anion were carried out at high temperature (*53-55*). In these systems the solvation was found to occur on two timescales, both of which were dependent upon cation size and slower than that observed for molecular liquids. Very recently, the first study on an ammonium room-temperature ionic liquid was reported (*44*). A rapid relaxation component for methyltributylammonium [NTf$_2$]$^-$ was not observed; the solvation response was highly non-exponential and was fitted to a single stretched exponential with lifetime 1.3 ns and $\beta = 0.39$, giving an average lifetime of 4.5 ± 1.5 ns (viscosity 520 cP). This average lifetime is consistent with the ~ 4 ns solvation time estimated for the electron in this liquid (*7*). Coumarin-153 Stokes shift measurements indicate that the average solvation lifetime in highly viscous [NBu$_3$Hx][CH$_3$SO$_3$] (12900 cP at 20 °C) is 86 ns (*56*). The fast solvation component was also not observed in phosphonium-based ionic liquids (*44, 57*), giving rise to the suggestion that the fast component observed in the imidazolium cases is due to local rearrangements of coplanar C153-imidazolium cation complexes (*44*).

The rotational dynamics of solvents may be probed by employing fluorescence polarization anisotropy measurements. Whilst these are also essential to the understanding of the overall dynamics of solvents, to date very little data of this type has been reported. If hydrodynamic theory holds for this new class of solvents, the rotation time should be proportional to viscosity and inversely proportional to temperature. The kinetics of rotation of the imidazolium ionic liquid butylmethylimidazolium [PF$_6$]$^-$ can be described by a single exponential anisotropy decay, and is found to conform to hydrodynamic theory (*43*). Recent experiments on ammonium [NTf$_2$]$^-$ salts (*56*) and phosphonium salts of several anions (*57*) indicate that rotational anisotropy decay occurs on multiple timescales, typical of glassy systems and most likely due to spatial heterogeneity of the chromophore solvation environment.

Static and Dynamic Solvation of Excess Electrons in Ionic Liquids

Electron solvation in molecular solvents has been intensively studied since the discovery of the ammoniacal electron (*58, 59*) and observations of the aqueous electron by Hart (*60, 61*). Solvated electron spectra have been measured in a wide variety of liquids, primarily with radiolysis techniques due to the transient nature of the excess electron in most solvents (*62-64*). The observed spectra result from transitions from the equilibrated electronic ground state to a manifold of bound and continuum excited states. The energies of these transitions depend on the strength of the solvation interaction and the size, shape and symmetry of the cavity formed in the solvent by the excess electron. The effect of the solvent on e_{solv}^- spectra can be seen in Figure 4, in which the absorption spectrum of e_{solv}^- in [NBu₃Me][NTf₂]⁻ (7) is shown in relation to those found in representative molecular solvents. It is interesting to note that the peak position is comparable to those found in diamines or acetonitrile (not shown); given the ionic environment, intuition would have expected a bluer transition. Molecular dynamics simulations of charged cavities in ionic liquids will be needed to interpret such spectra.

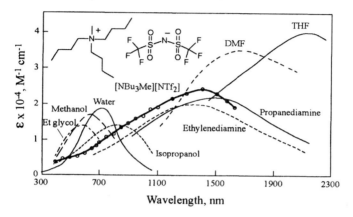

Figure 4. Absorption spectra of the solvated electron in representative molecular liquids (data from ref. 57) and [NBu₃Me][NTf₂] (ref. 7).

The sensitivity of the solvated electron spectrum to relatively subtle structural variation is illustrated in Figure 5, which shows the e_{solv}^- spectra observed in [C₄mpyrr][NTf₂] and [NBu₃Hx][NTf₂] (26). Each spectrum is representative of several liquids with similar structures.

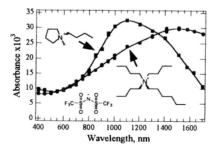

Figure 5. Transient absorbance spectra of the solvated electron in methylbutyl-pyrrolidinium [NTf₂] (squares) and hexyltributylammonium [NTf₂] (circles).

112

The liquids with alkyl- or ether-pyrrolidinium cations have e_{solv}^- absorbance maxima around 1100 nm while tetraalkylammonium salts have maxima in the region of 1400-1500 nm. The differences in band shape and position indicate that different distributions of electron solvation cavity sizes and shapes result as a function of cation structure. It may be possible to use molecular dynamics simulations to model the structure-induced cavity variations between the two families of cations.

Ionic liquids can be modified with specific prosthetic groups for the purpose of examining their effects on solvation dynamics. An example is the series of dicationic liquids of the type $[(CH_3)_2(R'')N(CH_2)_nN(R'')(CH_3)_2][NTf_2]_2$, where $R'' = (CH_2)_3OH$, $(CH_2)_2OCH_2CH_3$, or $(CH_2)_3CH_3$ and n = 3–8 (65). Solvated electrons in solvents with hydroxyl groups (such as alcohols and water) typically have absorption maxima around 600 – 700 nm, whereas the absorption maxima in the above ionic liquids containing ether or alkyl groups occur in the range of 1000 – 1100 nm. In the liquids where $R'' = (CH_2)_3OH$, rearrangement of the solvation environment from alkane-like to alcohol-like for a sub-population of electrons is observed to occur with time constants of 25 – 40 ns (65). An example is shown in Figure 6 for n = 4.

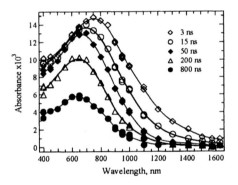

Figure 6. Pulse radiolysis transient absorption spectra at indicated time intervals for[(CH₃)₂((CH₂)₃OH)N(CH₂)₄N((CH₂)₃OH)(CH₃)₂][NTf₂]₂.

Summary

This chapter highlights early studies into solvation phenomena in non-imidazolium ionic liquids and outlines directions for further work. The reactivity of transient species such as the excess electron will be the major determinant of the radiation stability of ionic liquids for important applications in processing nuclear fuel and waste. Solvation of electrons and the consequent effects on their energetics and mobility influence reactivity patterns. Understanding ionic liquid solvent response to charge movement is key to design of charge transport systems to effect photochemical and photoelectrochemical solar energy conversion. Work to date has established tentative connections between solvation dynamics measured with fluorescent dyes and absorption spectroscopy of electron solvation. In order to make detailed correlations it has been necessary to develop ionic liquids that are better suited to the experimental requirements of ultrafast pulse radiolysis studies. These new tools will permit extensive measurement of electron solvation on picosecond timescales. Other liquids have been designed to control the primary radiation chemistry to produce oxidizing radicals ($[Br_2]^{\cdot-}$) or H-atoms for studies of diffusion rates. The foundation has been laid for using pulse radiolysis to explore charge transport phenomena in ionic liquids, and to interpret the observations using detailed solvation dynamics information presently being obtained for a wide range of compositions and viscosities.

Acknowledgments

The work was performed at Brookhaven National Laboratory under contract DE-AC02-98CH10886 with the U.S. Department of Energy and supported by its Division of Chemical Sciences, Office of Basic Energy Sciences and the BNL Laboratory Directed Research and Development Program. The authors would like to thank Pedatsur Neta, Edward Castner, Sharon Lall-Ramnarine, Robert Engel, Tomasz Szreder, Mark Kobrak, Elina Trofimovsky, Kimberly Odynocki and R. Dave Ramkirath for their collaborations on ionic liquids. The authors thank Cytec Canada Inc. for supplying samples of phosphonium ionic liquids.

References

1. Harmon, C. D., Smith, W. H., Costa, D. A., *Radiat. Phys. Chem.,* **2001**, *60,* 157-159.
2. Allen, D., Baston, G., Bradley, A. E., Gorman, T., Haile, A., Hamblett, I., Hatter, J. E., Healey, M. J. F., Hodgson, B., Lewin, R., Lovell, K. V., Newton, B., Pitner, W. R., Rooney, D. W., Sanders, D., Seddon, K. R., Sims, H. E., Thied, R. C., *Green Chemistry,* **2002,** *4,* 152-158.
3. Wishart, J. F., Nocera, D. G., eds. *Photochemistry and Radiation Chemistry. Complementary Methods for the Study of Electron Transfer.* Advances in Chemistry Series. Vol. 254. 1998, American Chemical Society: Washington, DC.

114

4. Wishart, J. F., *Accelerators for Ultrafast Phenomena*, in *Radiation Chemistry: Present Status and Future Trends*, Jonah, C. D. and Rao, B. S. M., Editors. **2001**, Elsevier Science: Amsterdam. p. 21-35.
5. Wishart, J. F., Cook, A. R., *Rev. Sci. Instr.*, in press.
6. Wishart, J. F., *Radiation Chemistry of Ionic Liquids: Reactivity of Primary Species*, in *Ionic Liquids as Green Solvents. Progress and Prospects*, Rogers, R. D. and Seddon, K. R., Editors. **2003**, American Chemical Society: Washington, DC. p. 381-396.
7. Wishart, J. F., Neta, P., *J. Phys. Chem. B*, **2003**, *107*, 7261-7267.
8. Grodkowski, J., Neta, P., Wishart, J. F., *J. Phys. Chem. A*, **2003**, *107*, 9794-9799.
9. Grodkowski, J., Neta, P., *J. Phys. Chem. A*, **2002**, *106*, 5468-5473.
10. Grodkowski, J., Neta, P., *J. Phys. Chem. A*, **2002**, *106*, 9030-9035.
11. Behar, D., Neta, P., Schultheisz, C., *J. Phys. Chem. A*, **2002**, *106*, 3139-3147.
12. Skrzypczak, A., Neta, P., *J. Phys. Chem. A*, **2003**, *107*, 7800-7803.
13. Grodkowski, J., Neta, P., *J. Phys. Chem. A*, **2002**, *106*, 11130-11134.
14. McLean, A. J., Muldoon, M. J., Gordon, C. M., Dunkin, I. R., *Chem. Comm.*, **2002**, 1880-1881.
15. Debye, P., *Trans. Electrochem. Soc.*, **1942**, *82*, 265-272.
16. Backstrom, H. L. J., Sandros, K., *Acta. Chem. Scand.*, **1960**, *14*, 48-62.
17. Behar, D., Gonzalez, C., Neta, P., *J. Phys. Chem. A*, **2001**, *105*, 7607-7614.
18. Matsumoto, H., Kageyama, H., Miyazaki, Y., *Chem. Lett.*, **2001**, *2*, 182-183.
19. Matsumoto, H., Kageyama, H., Miyazaki, Y., *Chem. Comm.*, **2002**, 1726-1727.
20. MacFarlane, D. R., Meakin, P., Sun, J., Amini, N., Forsyth, M., *J. Phys. Chem. B*, **1999**, *103*, 4164-4170.
21. MacFarlane, D. R., Sun, J., Golding, J., Meakin, P., Forsyth, M., *Electrochimica Acta*, **2000**, *45*, 1271-1278.
22. Cooper, E. I., Angell, C. A., *Solid State Ionics*, **1983**, *9 & 10*, 617-622.
23. Cooper, E. I., Angell, C. A., *Solid State Ionics*, **1986**, *18 & 19*, 570-576.
24. Matsumoto, H., Yanagida, M., Tanimoto, K., Nomura, M., Kitagawa, Y., Miyazaki, Y., *Chem. Lett.*, **2000**, *8*, 922-923.
25. Wasserscheid, P., Hal, R. v., Bosmann, A., Esser, J., Jess, A., in *Ionic Liquids as Green Solvents. Progress and Prospects*, Rogers, R. D. and Seddon, K. R., Editors. **2003**, American Chemical Society: Washington, DC. p. 57-69.
26. Wishart, J. F., Funston, A. M., *et al, Unpublished Data*.
27. Golding, J., Forsyth, S., MacFarlane, D. R., Forsyth, M., Deacon, G. B., *Green Chemistry*, **2002**, *4*, 223-229.
28. Golding, J., Hamid, N., MacFarlane, D. R., Forsyth, M., Forsyth, C., Collins, C., Huang, J., *Chem. Mater.*, **2001**, *13*, 558-564.
29. MacFarlane, D. R., Golding, J., Forsyth, S., Forsyth, M., Deakin, G. B., *Chem. Comm.*, **2001**, 1430-1431.

30. Sun, J., Forsyth, M., MacFarlane, D. R., *J. Phys. Chem. B*, **1998**, *102*, 8858-8864.
31. Forsyth, C. M., MacFarlane, D. R., Golding, J. J., Huang, J., Sun, J., Forsyth, M., *Chem. Mater.*, **2002**, *14*, 2103-2108.
32. Susan, M. A. B. H., Noda, A., Mitsushima, S., Watanabe, M., *Chem. Comm.*, **2003**, 938-939.
33. Wasserchied, P., Welton, T., eds. *Ionic Liquids in Synthesis*. 2003, Wiley-VCH: Weinheim.
34. Xu, W., Cooper, E. I., Angell, C. A., *J. Phys. Chem. B*, **2003**, *107*, 6170-6178.
35. Angell, C. A., Xu, W., Yoshizawa, M., Belieres, J.-P., *International Symposium on Ionic Liquids in Honour of Marcelle Gaune-Escard*, **2003**.
36. Xu, W., Angell, C. A., *Science*, **2003**, *302*, 422-425.
37. Yoshizawa, M., Xu, W., Angell, C. A., *J. Am. Chem. Soc.*, **2003**, *125*, 15411-15419.
38. Pringle, J. M., Golding, J., Baranyai, K., Forsyth, C. M., Deakin, G. B., Scott, J. L., MacFarlane, D. R., *New. J. Chem.*, **2003**, *27*, 1504-1510.
39. Cave, R. J., Castner, E. J., *J. Phys. Chem. A*, **2002**, *106*, 12117-12123.
40. McCarthy, P. K., Blanchard, G. J., *J. Phys. Chem.*, **1993**, *97*, 12205-12209.
41. Lewis, J. E., Maroncelli, M., *Chem. Phys. Lett.*, **1998**, *282*, 197-203.
42. Jiang, Y., McCarthy, P. K., Blanchard, G. J., *Chem. Phys.*, **1994**, *183*, 249-267.
43. Ingram, J. A., Moog, R. S., Ito, N., Biswas, R., Maroncelli, M., *J. Phys. Chem. B*, **2003**, *107*, 5926-5932.
44. Arzhantsev, S., Ito, N., Heitz, M., Maroncelli, M., *Chem. Phys. Lett.*, **2003**, *381*, 278-286.
45. Chakrabarty, D., Hazra, P., Chakraborty, A., Seth, D., Sarkar, N., *Chem. Phys. Lett.*, **2003**, *381*, 697-704.
46. Karmakar, R., Samanta, A., *J. Phys. Chem. A*, **2003**, *107*, 7340-7346.
47. Karmakar, R., Samanta, A., *J. Phys. Chem. A*, **2002**, *106*, 6670-6675.
48. Karmakar, R., Samanta, A., *J. Phys. Chem. A*, **2002**, *106*, 4447-4452.
49. Fee, R. S., Maroncelli, M., *Chem. Phys.*, **1994**, *183*, 235-247.
50. Giruad, G., Gordon, C. M., Dunkin, I. R., Wynne, K., *J. Chem. Phys.*, **2003**, *119*, 464-476.
51. Hyun, B.-R., Dzyuba, S. V., Bartsch, R. A., Quitevis, E. L., *J. Phys. Chem. A*, **2002**, *106*, 7579-7585.
52. Cang, H., Li, J., Fayer, M. D., *J. Chem. Phys.*, **2003**, *119*, 13017-13023.
53. Bart, E., Meltsin, A., Huppert, D., *J. Phys. Chem.*, **1994**, *98*, 3295-3299.
54. Bart, E., Meltsin, A., Huppert, D., *J. Phys. Chem.*, **1994**, *98*, 10819-10823.
55. Bart, E., Meltsin, A., Huppert, D., *J. Phys. Chem.*, **1995**, *99*, 9253-9257.
56. Funston, A. M., Wishart, J. F., Bird, R., Grant, C., Castner, E. J., *Unpublished Data*.
57. Maroncelli, M., *Private communication*.
58. Belloni, J., Billiau, F., Saito, E., *Nouv. J. Chim.*, **1979**, *3*, 157-161.
59. Weyl, W., *Ann. Phys.*, **1863**, *197*, 601.

60. Hart, E. J., Boag, J. W., *J. Am. Chem. Soc.,* **1962**, *84,* 4090.
61. Nielsen, S. O., Michael, B. D., Hart, E. J., *J. Phys. Chem.,* **1976**, *80,* 2482-2488.
62. Dorfman, L. M., Galvas, J. F., in *Radiation Research. Biomedical, Chemical and Physical Perspectives*, Nygaard, O. F., Adler, H. J., and Sinclair, W. K., Editors. **1975**, Academic Press: New York. p. 326-332.
63. Jay-Gerin, J.-P., Ferradini, C., *Can. J. Chem.,* **1990**, *68,* 553-557.
64. Belloni, J., Marignier, J. L., *Radiat. Phys. Chem.,* **1989**, *34,* 157-171.
65. Wishart, J. F., Lall-Ramnarine, S. I., Raju, R., Scumpia, A., Bellevue, S., Ragbir, R., Engel R., *Radiat. Phys. Chem.,* in press.

Theory and Modeling

Chapter 9

Force Field Development and Liquid State Simulations on Ionic Liquids

Jones de Andrade, Elvis S. Böes, and Hubert Stassen*

Grupo de Química Teórica, Instituto de Química, Universidade Federal do Rio Grande do Sul, 91.540–000 Porto Alegre-RS, Brazil
*Corresponding author: gullit@iq.ufrgs.br

Performing quantum mechanical computations and molecular mechanics calculations, an all-atom force field compatible with the AMBER methodology has been developed for ionic liquids containing 1,3-dialkylimidazolium cations. The force field was employed in extensive molecular dynamics simulations on the condensed phase of several ionic liquids. The influence of the alkyl chain length and the anion size on equilibrium properties has been investigated. We present results for the densities, internal energies, and structural features in terms of radial and angular distribution functions for several ionic liquids based on the 1,3-dialkylimidazolium cation. In addition, the diffusion coefficients and electric conductivities were computed for these ionic liquids. The good agreement between the computed data and experimental findings demonstrates that the obtained force field is able to represent the liquid phase of the considered ionic liquids.

A Historical Brieffing...

The Room Temperature Molten Salts, now broadly accepted as Ionic Liquids (ILs) were first reported in 1914 (*1*). The ethyl-pyridinium cations were intensively studied by 1948 (*2*). The butyl-pyridinium were highlighted in the 1970s (*3*). The imidazolium cation class was developed in 1980s (*4*), while moisture stable ionic liquids first appeared in the last decade (*5, 6*).

A few quantum mechanical (QM) studies are reported in the literature. We mention the semiempirical studies concerning the stability (which led to the discovery of the whole dialkylimidazolium class of cations) (*7*) and the cation-anion pair structure (*8*). Several *abInitio* studies were reported, focusing on spectroscopic properties of the ions (*9*), structural features related to X-ray liquid phase diffraction measurements (*10, 11, 12*) and hydrogen bonding (*13*).

Simulations involving ILs in its liquid phase were initiated in 2000 (*13*). First, both the United-Atom (UA) and All-Atom (AA) rigid molecule approach, combined with Buckingham Van-der-Waals potentials and charges from DMA model (*13*), were developed by Lynden-Bell and Hanke (*13*) for the 1,3-Dimethylimidazolium (MMI^+) and 1-Ethyl-3-Methylimidazolium (EMI^+) cations and the chloride (Cl^-) and hexafluorophosphate (PF_6^-) anions. This Force-Field (FF) has also been employed in calculations of MMI·Cl/Benzene mixtures (*14*) and of chemical potentials for many simple molecules dissolved in ILs (*15, 16*).

A second group of approaches, based on the flexible molecule OPLS collection of FFs (*17*), was proposed by Maginn and Brennecke (*18*). They focused their studies on the IL composed by 1-*n*-buthyl-3-methylimidazolium cations (BMI^+) and the PF_6^- anions. They have also developed two versions of the FF, an UA and an AA, with CHELPG type of charges (*19*), calculated at RHF/6-31G(d) level of theory. These calculations were directed towards many physical properties and structural data using Monte-Carlo (MC) techniques. Another FF, also based on OPLS-AA (*17*), was developed by Berne and Margulis (*20*), employing the ESP (intead of CHELPG) scheme (*21*) for the calculation of atomic charges.

A third approach based on the CHARMm22 FF (*22*), is an AA FF that uses CHELPG type charges (*19*). Also developed by Maginn and Brennecke (*23*), it has been applied to the BMI·PF_6 IL, mainly in calculation of physical properties of this IL.

The last type of FF is the one developed by our group. It is an approach based on the AMBER FF (*24*), RESP charges (*25*) calculated at UHF/6-31G(d) level of theory. In its development, we made use of analogies with the protonated hystidine aminoacid, whose residue possesses the imidazolium ring from the ILs (*26, 27*). This FF is of AA type, fully flexible and easily extendable to other ILs of the dialkylimidazolium class. Up to now, it has been applied in the studies of aluminium-tetrachloride ($AlCl_4^-$) and boron-tetrafluoride (BF_4^-) anions, combined with the MMI⁺, EMI⁺, 1-*n*-propyl-3-methylimidazolium (PMI⁺) and BMI⁺ cations. Here, many of our calculations are presented, ranging from QM to Molecular Dynamics (MD) techniques. Physical and physical-chemical properties obtained from these calculations are described.

From Quantum, to Classical

Nowadays, the computational treatise of complex liquid systems still depends on classical simulation methods. Thus, it is necessary to model a system that behaves in the best way possible, like the "Real World".

One way to do that would be to develop an entire FF using and validating it against experimental data. Unfortunatelly, besides the inherent difficulties that would arise from such a procedure, there is an inimaginable lack of specific data. For instance, there are only few experimental data for the liquid structure of ILs, almost no data for the self-diffusion constant for the ILs ions and no experimental data for the vaporization enthalpy of the ILs.

The lack of these experimental information by directly using the "Real World" values retains us. A consistent alternative is represented by the FF development from first principles, which is an "enougthly good aproximation for the Real World", via QM methodologies.

The Bond Orders: Main Clue for the Choice of Bond Parameters

In fact, the analogy between imidazolium cations and the protonated hystidine aminoacid represents an initial approach to model the imidazolium ring of the cations. Modern FFs generally obtain a set of carefully chosen potential parameters for aminoacids, adopting the hystidine analogy, and we reached at a point where many of the intramolecular parameters for the cations, like those involving the alkyl chains, could easily be chosen. On the other hand, other parameter selections were less obvious.

This is mainly related to the atoms of the imidazolium ring. At a glance, the ring follows the Hückel rules of aromaticity. However, the question that arises is: To what extent is the imidazolium ring aromatic? And: What is the most probable distribution of single and double bonds in the ring?

To get some answers, we decided to perform QM calculations for the cations, aiming at the computationally unexpensive bond orders. In this way, one can compare bond orders for the imidazolium ring system and others used as standards, like benzene, ethylene, N,N-dimethylamine and, especially, imidazole. The results of these calculations are summarized in Table I.

The calculations lead to an unexpected conclusion: although there is some degree of deslocalization of π electron density within the ring, it is not really aromatic. This leads to and supports some interesting approaches employed in here, like considering the C4 and C5 atoms of the imidazolium ring as single sp^2 hybridized *non*-aromatic atoms, with the N1-C5 and N3-C4 bonds as simple, whereas there is a considerable deslocalization of π electron density between the N1-C2 and N3-C2 bonds. This is more clearly seen when comparing the imidazolium ring bond orders with the imidazole bond orders.

Table I. Bond lengths and orders from the imidazolium ring and example molecules (calculated at HF/6-31G(d) level of theory).

Molecule	Bond	Bond Length (Å)	Bond Order
MMI$^+$	N1-C2	1.316	1.304
	N3-C4	1.378	1.057
	C4-C5	1.342	1.762
Benzene	C-C	1.386	1.454
	C-H	1.076	0.942
Imidazole	N1$_{(H)}$-C2	1.349	1.117
	C2-N3	1.289	1.622
	N3-C4	1.372	1.232
	C4-C5	1.350	1.688
	C5-N1$_{(H)}$	1.372	1.047
Ethylene	C-C	1.317	1.983
	C-H	1.076	0.956
N,N-dimethylamine	C-H	1.085	0.958
	C-N	1.447	0.939
	N-H	1.001	0.851

Using these conclusions, the force field can be described as a proper combination of parameters that come from two definitions of atoms for the dialkylimidazolium cations, defined as depicted in figures 1a and 1b (in both, the atom types with the nomenclature from AMBER94 (*24*) are shown). The first figure (1a) shows the atoms using the analogy with the protonated hystidine aminoacid residue; In the second figure (1b), used to complete the lacking parameters, the best choice of atoms from the bond order QM calculations is illustrated.

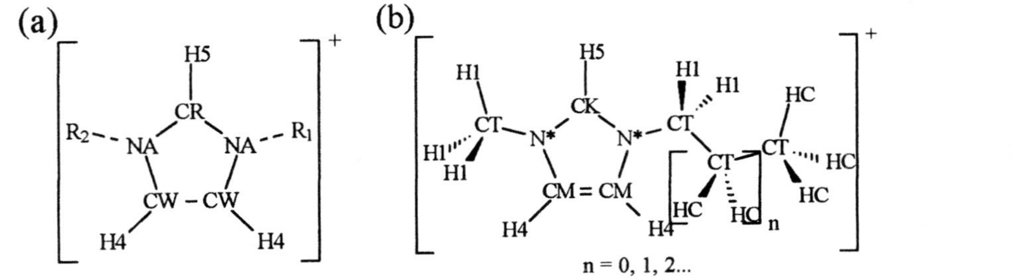

Figure 1. (a) Choice of atom types considering analogies with the protonated hystidine residue using the AMBER94 nomenclature; (b) Choice of atom types, from the bond order analysis using the AMBER94 nomenclature.

Atomic Charges: Modelling the Main Interaction in Ionic Liquids

Since an Ionic Liquid is basically composed by independent and oppositely charged molecules, it appears to be obvious that coulombic interactions play an important role. Thus, an appropriate modeling of the charges represents a very important aspect in the FF development for ILs.

The approach that is widely used, especially in the case of ILs, is to compute the atomic point charges from QM calculations. Different charge type models have already been used and reported in the literature: DMA (*13*) in the Lynden-Bell *et al.* FFs (*13*); ESP (*21*) for Berne *et al.* (*20*), CHELPG (*19*) for Maggin (*18, 23*) and, in the FF here reported (*26, 27*), we use the RESP model (*25*) for the charges.

This type of charges was especially developed for the use with the AMBER94 FF (*24*). This leaded to some kind of "dependent parametrization" in the AMBER 94 FF for some of the flexible potentials employed, especialy the dihedral torsions.

In the RESP model, charges are optimized to values that better reproduce the overall electrostatic surrounding of the molecule, as obtained from the QM calculation. The model leads to interestingly different atomic charges than in the other AA FFs available in the literature. For comparison, these charges for the imidazolium atoms in two cations are summarized in Table 2. The fact that must be strongly stressed here is that the RESP charges (*25*) produce imidazolium hydrogens that are more acid than in the other models.

In order to analyze the performance of the proposed charge model in reproducing the QM electrostatic surrounding for the cations, the molecular dipole moments of the cations (from both the classical point charges and the QM calculations) are compared. These data, shown in Table III, prove the efficiency of the RESP model in order to reproduce the electrostatic behavior of the molecular cation.

Table II. Imidazolium ring atomic charges (in atomic units) from different models reported in the literature.

Imidazolium Ring Atoms*	EMI^+		BMI^+		
	RESP	DMA	RESP	CHELPG	ESP
N1	-0.010004	-0.267	+0.045614	+0.111	+0.06881
C2	+0.058110	+0.407	+0.007624	+0.056	-0.05841
N3	+0.079453	-0.267	+0.061445	+0.133	+0.24016
C4	-0.167115	+0.105	-0.126156	-0.141	-0.27761
C5	-0.192460	+0.105	-0.216900	-0.217	-0.12324
H2	+0.204594	+0.097	+0.230517	+0.177	+0.23760
H4	+0.248174	+0.094	+0.231249	+0.181	+0.25557
H5	+0.258666	+0.094	+0.269228	+0.207	+0.24141

*Atoms numbering within IUPAC standards.

Table III. Dipole Moments from Quantum Mechanics (with UHF/6-31G(d) basis set) and the the modelled charges for MMI^+, EMI^+, PMI^+ and BMI^+.

Cation	UHF/6-31G(d)	RESP Charges
MMI^+	1.078692D	1.08711D
EMI^+	1.761863D	1.78362D
PMI^+	3.493592D	3.51203D
BMI^+	5.563889D	5.60577D

Optimized Geometry and Frequencies: The Mechanical Validation

There are two basic properties of the modelled molecule (or, in this case, ions) that the FF should reproduce for being considered acceptable: the molegular geometry corresponding to minimized energy of the molecule and its flexibility.

The ability of the FF proposal to reproduce the ion's geometry can be tested against: experimental data, usually from X-ray diffraction measurements (28); and theoretical structures, from the QM geometry optimization. The FF is tested by a Molecular Mechanics (MM) energy minimization, which must reproduce in the best possible way the geometry of the molecule from the other two approaches. Doing so, one can compare bond lengths, bond angles and dihedral angles of the molecule in all three approaches. In cases of deviations, such an analysis also provides usefull information about the deficiencies in the FF and guidelines for their improvement (27).

Another property of the molecule that should be reproduced by the FF is the flexibility. Again, experimental data from vibrational spectroscopy like RAMAN (9) or IR techniques (29), or theoretical procedures like a Normal Mode Analysis (NMA) performed on QM level serve as a guideline. The developed FF is applied to a MM/NMA using the FF, and the normal modes of vibration can be compared with the other two approaches (27). As before, this kind of analysis facilitates a FF correction and the validation of its flexibility.

At this stage, the FF is already proposed and validated in terms of MM. The next step is to observe its performance in liquid phase simulations.

From Mechanics, to Dynamics

In order to carry out simulations on ILs described by a developed FF, one must perform its validation against properties for the bulk liquid. This can be achieved by different simulation techniques, like MC, Stochastic Dynamics or, as we employed, the Molecular Dynamics (MD) methodology (30, 31).

In this method, classical equations of motion are numerically resolved using the model potential presented and validated before. Doing so, it is possible to obtain a realistic liquid phase configuration and to follow its evolution in time, in such a way that one can average many properties of the bulk liquid. This technique is already standard and very well described in the literature (30, 31).

Molecular Geometries in the Liquid Phase, Energies, Densities...

Molecular geometry parameters in the liquid phase and interaction energies represent examples of averages that are readily available from a MD simulation. The first can be used to compare with the crystal structure available from X-ray diffraction measurements. Thus, one might observe differences and similarities in some configurational features between the liquid and the solid phases (27).

$$\Delta H_{vap} = -U + RT \qquad (1)$$

Summing the intermolecular pair interaction energies from the FF proposal the internal energy of the liquid (U) can be acquired. This can be related to the enthalpy of vaporization by the equation (1). Results for several ILs are summarized in table IV, where one might notice that the *non*-volatil character of the ILs is well reproduced by the strongly negative interaction energies and large entalphies of vaporization from the simulations (27).

Another property that can be employed to validate the FF is the liquid density, which can be averaged from a simulation using a constant pressure and temperature (NpT) ensemble (27). Computed densities are summarized in table IV. The comparison with experimental densities shows that our FF successfully reproduces the experimental data with a very good level of accuracy. Moreover, these data indicate that the intermolecular potential parameters for both, the Van der Waals and coulombic interactions, are accurately estimated.

Table IV. Simulated internal energy (U), enthapy of vaporization (ΔH_{vap}) and density ($\rho_{simu.}$), and experimental density ($\rho_{exp.}$).

IL	U (kJ/mol)	ΔH_{vap} (kJ/mol)	$\rho_{simu.}$ (g/cm^3)	$\rho_{exp.}$ (g/cm^3)
MMI·AlCl$_4$ [a]	-285.134	288.152	1.3529	1.329
EMI·AlCl$_4$ [b]	-304.577	307.055	1.3159	1.3020
EMI·BF$_4$ [c]	-322.568	325.021	1.2547	1.24
PMI·AlCl$_4$ [b]	-306.880	309.358	1.3182	1.2624
BMI·AlCl$_4$ [b]	-325.291	327.769	1.2294	1.238
BMI·BF$_4$ [d]	-339.821	342.340	1.1738	1.17

a: simulated at 363K; b: simulated at 298K; c: simulated at 295K; d: simulated at 303K.

The Transport Coefficients

A good FF proposal for liquid state simulation is also required to pursuit adequate dynamical properties. In the case of ILs, a lot of experimental viscosities and electrical conductivity are available in the literature. However, the computational approach to these data is very expensive. On the other side, the computationally most simple and usefull self-diffusion coefficients of the ions in the ILs, is experimentally complicated and rare to find.

The computational difficulties that must be overcome to properly acquire the viscosity and electrical conductivities arise from the large statistical requirements to sample these quantities from equilibrium MD simulations via Green-Kubo (GK) time correlation functions. The self-diffusion coefficientes, however, can be easilly calculated from the slope of the mean square displacement, defined as in equation 2 (30, 31). Applied to ILs, equation (2) should be averaged over many independent initial configurations in order to prevent statistical inaccuracies related to the restricted motion of cation-anion.

$$D = \frac{1}{6t}\left\langle \left[r(t_0 + t) - r(t_0)\right]^2 \right\rangle \qquad (2)$$

Besides the fact that measuring the viscosities and electrical conductivities from computer simulations is quite difficult, we have performed qualitative comparisons of the experimental transport coefficients for viscosity and electrical conducitivity, and the average calculated self-diffusion coefficients. The results are shown in the figures 2a and 2b respectivelly. They show a qualitative agreement, in such a way that the electrical conductivity increases with the diffusion, while the viscosity decreases with it.

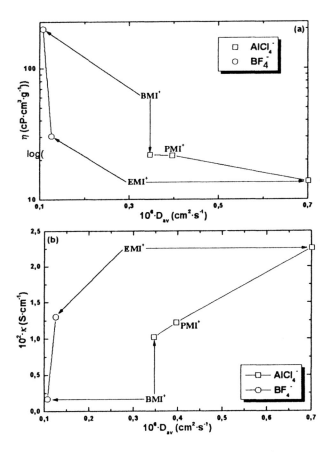

Figure 2. (a) Experimental kinematic viscosity (η_{exp}, in logaritimic scale) against the averaged self-diffusion coefficient of the cations and anions (D_{av}) from simulation; (b) Experimental electrical conductivity (κ_{exp}, in linear scale) against the averaged self-diffusion coefficient of the cations and anions (D_{av}) from simulation.

From the Moving Ions, to the Liquid Structure

Since the FF is reliable in reproducing the experimental features of the ILs, we come up with a discussion about the liquid phase structure. Here, we present results for the radial distribution funtions (RDFs) and the spatial distribution functions (SDFs).

How to Validate a Structure?

Although there is growing interest in the liquid ILs structure, up to now, only few studies concerning experimental ILs structures have been published. One might cite Hardacre *et al.* (*32, 33*), and also the investigations from Kohara and Takhashi (*10, 11, 12*) treating the long range structure of ILs (in terms of RDFs and structure factors S(q)). Other measures, including ^{1}H-NMR, are important only for a qualitative short range distribution analysis.

The RDF for distances r_{ij} between atoms i and j belonging to different ions is defined by equation (3).

$$g(r) = \frac{1}{\rho \, \Delta V_{r_n}} \left\langle \sum_i \sum_{j>i} \int_{r_n}^{r_{n+1}} \delta \left(r - r_{ij}\right) dr \right\rangle \tag{3}$$

Once this total RDF is obtained, one can relate it to the experimental structure factor S(q) by fourier transform (FT). In figure 3, we compare the experimental $g_{neu}(r)$ for liquid EMI·AlCl$_4$ to the simulated curves from equation (4). Also included is a FT smothed from the simulated $g_{neu}(r)$ (*26, 27*).

To compare with experimental X-ray or neutron diffraction measurement, the averaged atom-atom RDFs defined by equation (3) must be wheightened in the proper way using the correct cross section factors (b_i). They are easilly combined with their atomic molar fractions in the liquid phase by the equation (4).

$$g_{neu}(r) = \frac{\sum_{ij} x_i x_j \overline{b_i b_j} g_{ij}(r)}{\left(\sum_i x_i b_i\right)^2} \tag{4}$$

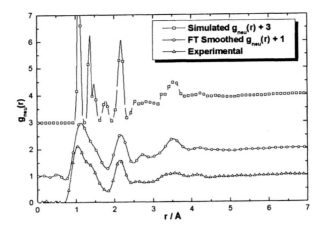

Figure 3. Experimental and neutron weightened RDFs (direct and FT smoothed results).

From this figure, it becomes evident that the simulation provides a reliable liquid phase structure for the ILs. Thus, we might get a deeper insight into the ILs liquid structure by simulations using the FF proposed here. We highlight that, to our knowledge, our FF proposal is the only one that has been validated up to now by this kind of comparison with experimental structural data for the ILs.

Hydrogen Bonds or Cation-Anion Interactions?

One of the most interesting structural feature of the ILs is related to the existence of hydrogen bonds, between the imidazolium ring hydrogens and the anions. These can easily stabilize a macro long-range organization of the liquid, and many of the ILs properties are supposed to be related to these hydrogen bonds.

In order to have a microscopical insight, we concentrate on some particular RDFs from equation (3) and somes SDFs defined by equation (5).

$$g(r,\theta) = \frac{1}{\rho \, \Delta V_{r_n,\theta_m}} \left\langle \sum_i \sum_{j>i} \int_{r_n}^{r_{n+1}} \delta(r - r_{ij}) dr \int_{\theta_m}^{\theta_{m+1}} \delta(\theta - \theta_{\hat{ijk}}) d\theta \right\rangle \qquad (5)$$

Hydrogen bonding might be identified from the RDFs for distances between the imidazolium ring hydrogens and the donator atoms of the anions. These RDFs are depicted (only for the hydrogen H2 from the imidazolium ring) in figure 4. We observe a strong short-range peak at distances below 3Å indicating the formation of hydrogen bonds. On the other hand, this distances are slightly larger than common hydrogen bonds between hydrogen and chlorine or fluorine atoms.

In addition, some QM studies (at RHF/6-31G+(d,p) level of theory) of the cation-anion interactions were performed, trying to clarify whether the first peaks are related to a hydrogen bond or just cation-anion electrostatic interaction. Thus, QM calculations of cation–Cl$^-$ pairs have been performed. The distance, position, and bond order of the Cl$^-$ anion with respect to the H2 atom of the imidazolium ring has been computed. These data are compared with some model systems (water–Cl$^-$, Na$^+$–Cl$^-$) in Table V and indicate a strong ionic character in the H2–Cl interactions.

A particularly interesting SDF is represented by the angular distribution of the anion's donor atoms around the imidazolium C2-H2 bond. Defining the geometric center between N1 and N3 atoms as the origin of the coordinate system, this SDF is defined by the length of the position vector for the ation's donor atoms and the angle between that vector and the C2-H2 bond. In figure 5, the contour plots of this SDF is illustrated. The contour lines demonstrate that regions surrounding the imidazolium hydrogens are prefered positions for the anion's donor atoms. Moreover, comparing this SDF for MMI·AlCl$_4$ in figure 5a and BMI·AlCl$_4$ in figure 5b, one might conclude that the sterical effect due to longer alkyl chains implies reduced anion populations around the ring hydrogens.

Table V. QM results (at RHF/6-31G+(d,p) level of theory) for the hydrogen bonding behavior between the acceptor atoms and a Cl$^-$ anion.

Molecules	Acceptor Atom – Cl$^-$	
	Distance (Å)	Bond Order
EMI$^+$	2.113	0.241
BMI$^+$	2.129	0.228
1-n-hexyl-3-methylimidazolium	2.128	0.229
Water	2.401	0.083
Na$^+$	2.397	0.613

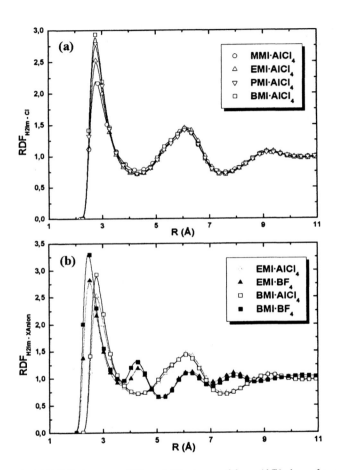

Figure 4. (a) RDFs between H2 and Cl atoms of four AlCl₄ based ions; (b) RDFs between H2 and the anion donor-atoms in EMI·AlCl₄, EMI·BF₄, BMI·AlCl₄ and BMI·BF₄.

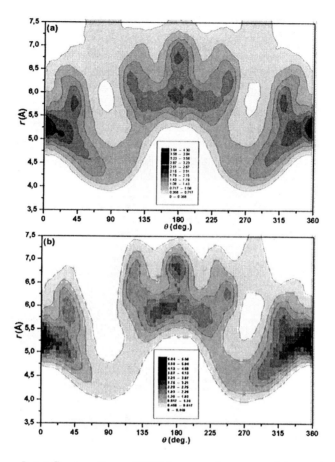

Figure 5. (a) Contour lines of SDF between the center of the imidazolium nitrogen atoms and the Cl atom of AlCl₄⁻ in MMI·AlCl₄; (b) Contour lines of SDF between the center of the imidazolium nitrogen atoms and the Cl atom of AlCl₄⁻ in BMI·AlCl₄.

Conclusions

In this study, we have shown that many of the experimental properties of ILs can be reproduced and then studied at a molecular level with use of the compulational and simulation techniques. These properties includes simple studies of the dipoles and molecular geometries of the ILs' cations, as well as

132

structural studies and dynamical properties. The computational treatment of ILs has just begun to be explored, and a lot of work still remains to be done in order to really clarify the experimental behavior of ILs, and furthermore, to push forward the development of new solvents within the Green Chemisty.

References

1. Walden, P. Bull. Acad. Imper. Sci. (St. Petersburg) **1914**, 1800.
2. Hurley, F. H. U. S. Patent 2.446.331, **1948**.
3. Chum, H. L.; Koch, V. R.; Miller, L. L.; Osteryoung, R. A. J. Am. Chem. Soc. **1975**, 97, 3264.
4. Boon, J. A.; Levisky, J. A.; Pflug, J. L.; Wilkes, J. S. *J. Org. Chem.* **1986**, *51*, 480.
5. Wilkes, J. S.; Zaworotko, M. J. J. Chem. Soc. Chem. Comm. **1992**, 965.
6. Fuller, J.; Carlin, R. T.; De Long, H. C.; Haworh, D. J. Chem Soc. Chem Comm. **1994**, 299.
7. Wilkes, J. S.; Levisky, J. A.; Wilson, R. A.; Hussey, C. L. Inorg. Chem. **1982**, 21, 1263.
8. Dymek Jr, C. J.;Stewart, J. J. P. *Inorg. Chem.* **1989**, *28*, 1472.
9. Takahashi, S.; Curtiss, L. A.; Gosztola, D.; Koura, N.; Saboungi, M.-L. *Inorg. Chem.* **1995**, *34*, 2990.
10. Takahashi, S.; Suzuya, K.; Kohara, S.; Koura, N.; Curtiss, L. A.; Saboungi, M.-L. *Z. Phys. Chem.* **1999**, *209*, 209.
11. Hagiwara, R.; Matsumoto, K.; Tsuda, T.; Ito, Y.; Kohara, S.; Suzuya, K.; Matsumoto, H.; Matsumoto, H. *J. Non-Cryst. Sol.* **2002**, *312*, 414.
12. Matsumoto, K.; Hagiwara, R.; Ito, Y.; Kohara, S.; Suzuya, K. *Nucl. Instr. Meth. Phys. Res. Section B-Beam Inter. Mater. Atoms* **2003**, *199*, 29.
13. Hanke, C. G.; S. L. Price; Lynden-Bell, R. M. *Mol. Phys.*, **2001**, *99*, 801.
14. Hanke, C. G.; Johansson, A.; Harper, J. B.; Lynden-Bell, R. M. *Chem. Phys. Lett.* **2003**, *374*, 85.
15. Hanke, C. G.; Atamas, N. A.; Lynden-Bell, R. M. *Green Chem.* **2002**, *4*, 107.
16. Lynden-Bell, R. M.; Atamas, N. A.; Vasilyuk, A.; Hanke, C. G. *Mol. Phys.* **2002**, *100*, 3225.
17. Jorgensen, W. L.; Maxwell, D. S.; Tirado-Rives, J. *J. Am. Chem. Soc.* **1996**, *118*, 11225.
18. Morrow, T. I.; Maginn, E. J. *J. Phys. Chem. B* **2002**, *106*, 12807.
19. Breneman, C. M.; Wibery, K. B. *J. Comput. Che.* **1990**, *11*, 361.
20. Margulis, C. J.; Stern, H. A.; Berne, B. J. *J. Phys. Chem. B* **2002**, *106*, 12107.
21. Chirian, L. E.; Francl, M. M. *J. Comput. Chem.* **1987**, *8*, 894.

22. MacKerell Jr., A. D.; Wiókiewicz-Kuczera, J.; Karplus, M. *J. Am. Chem. Soc,* **1995**, *117*, 11946.
23. Shah, J. K.; Brennecke, J. F.; Maginn, E. J. *Green Chem.* **2002**, *4*, 112.
24. Cornell, W. D.; Cieplak, P.; Bayly, C. I.; Gould, I. R.; Merz Jr, K. M.; Fergunson, D. M.; Spellmeyer, D. C.; Fox, T.; Caldwell, J. W.; Kollman, P. A. *J. Am. Chem. Soc.* **1995**, *117*, 5179.
25. Cornell, W. D.; Cieplak, P.; Bayly, C. I.; Kollman, P. A. *J. Am. Chem. Soc.* **1993**, *115*, 9620; and references therein
26. de Andrade, Jones; Böes, Elvis S.; Stassen, Hubert *J. Phys. Chem. B,* **2002**, *106*, p. 3546.
27. de Andrade, Jones; Böes, Elvis S.; Stassen, Hubert *J. Phys. Chem. B,* **2002**, *106*, p. 13344.
28. Elaiwi, A.; Hitchcock, P. B.; Seddon, K. R.; Srinivasan, N.; Tan, Y.-M.; Welton, T.; Zora, J. A. *J. Chem. Soc., Dalton* **1995**, 3467.
29. Campbell, J. L. E.; Johnson, K. E.; Torkel, J. R. *Inorg. Org.* **1994**, *33*, 3340.
30. Allen, M. P.; Tildesley, D. J. *Computer Simulation of Liquids*, 1st Ed; Oxford University Press Inc.: Oxford, 1997.
31. Smit, B.; Frenkel, D. *Understading Molecular Simulation*, 1st Ed.; Academic Press: San Diego, 1996.
32. Hardacre, C.; Holbrey, J. D.; McMath, S. E. J.; Bowron, D. T.; Soper, A. K. *J. Chem. Phys.* **2003**, *118*, 273.
33. Hardacre, C.; McMath, S. E. J.; Nieuwenhuyzen, M.; Bowron, D. T.; Soper, A. K. *J. Phys. Cond. Matt.* **2003**, *15*, S159.

Chapter 10

Modeling Ionic Liquids of the 1-Alkyl-3-methylimidazolium Family Using an All-Atom Force Field

J. N. Canongia Lopes[1], J. Deschamps[2], and A. A. H. Pádua[2]

[1]Centro de Química Estrutural, Instituto Superior Técnico,
Av. Rovisco Pais, 1049–001 Lisboa, Portugal
[2]Laboratoire de Thermodynamique des Solutions et des Polymères,
Université Blaise Pascal, 24 Av. des Landes, 63177 Aubière, France

A new force field for the molecular modeling of ionic liquids of the dialkylimidazolium cation family based on the OPLS-AA/AMBER framework is presented and discussed. *Ab initio* calculations were performed to obtain several terms in the force field not yet defined in the literature. These include torsion energy profiles and distributions of atomic charges that blend smoothly with the OPLS-AA specification for alkyl chains. Validation was carried out comparing simulated and experimental data on fourteen different salts, comprising three types of anion and five lengths of alkyl chain, both in the crystalline and liquid phases. The present model can be regarded as a step towards a general force field for ionic liquids of the imidazolium cation family that was built in a systematic way, is easily integrated with OPLS-AA/AMBER and is transferable between different combinations of cation-anion.

Introduction

Ionic liquids (ILs) have attracted much attention from the scientific community in recent years, their status as "green" or "designer" solvents explaining the large amount of studies concerning their possible industrial use as reaction or extraction media (*1*). Unfortunately, the number of studies regarding their thermophysical characterization is by no means so extensive and the amount and scope of molecular modeling work is also limited (*2–7*).

In this communication we introduce a new force field for ionic liquids (*7*) and review some of the currently available models (*2–6*). The comparison between force fields will focus on three main issues. i) Internal consistency. Many of the existing IL models borrow parameters from different, and not always compatible, sources. In this work *ab-initio* calculations are used extensively to provide essential data for the development of an internally consistent force field. This includes molecular geometry optimization and the description of electron density using extended basis sets, leading to the evaluation of force field parameters such as torsion energy profiles and point charges on the interaction centers. ii) Transferability. The parameterization of IL models should not be restricted to one molecular species since one wants to deal with families of similar compounds, generally the cations, that can be combined with different counter-ions. The proposed IL force field concentrates on the parameterization of parts of the molecules that are common to an entire family of cations and adopts a strategy to add specific substituents. The parameterization of the cations and anions is performed independently to ensure their interchangeability. iii) Compatibility. The model builds on the OPLS-AA force field (*8*). This means that any molecule or residue already defined in that database can be combined with the proposed force field. The choice of the OPLS-AA force field as a framework also benefits from its articulation with structural parameters from the even larger AMBER force field (*9*).

Force field development

One of the most extensively studied IL family has been that of the 1-alkyl-3-methyl-imidazolium cation with counter-ions such as PF_6^-, BF_4^- or Cl^- (Table I). The force fields used in most of the models discussed in this work take the functional form given in Table II. The different terms of the functional are discussed in the section of the manuscript given in the same table.

The parameterization of a force field requires the knowledge of two sets of molecular data corresponding to bonded and non-bonded interactions. The first set comprise internal molecular parameters that can be estimated from quantum mechanical calculations on isolated molecules and/or gas-phase spectroscopic data whereas the last set also includes external or intermolecular parameters to be inferred from auxiliary diffraction and/or condensed phase thermodynamic data. The nomenclature adopted in this work is given in Figure 1

Table I. Ionic Liquid models discussed in this work

Authors (ref.)	HPLB (2)	ABS (3)	SBM (4)	MM (5)	MSB (6)	CLDP (7)
Cations	$mmim^+$, $emim^+$	$emim^+$, $bmim^+$	$bmim^+$	$bmim^+$	$bmim^+$	$amim^+$ (C_1-C_{12})
Anions	Cl^-, PF_6^-	$AlCl_4^-$, BF_4^-	PF_6^-	PF_6^-	PF_6^-	PF_6^-, Cl^-, NO_3^-

Table II. Generic functional form of the force field. Subsection of the text where each term is discussed

Force field	Subsection
$$\sum_{bonds} K_r(r-r_{eq})^2 + \sum_{angles} K_\theta(\theta-\theta_{eq})^2 +$$	→ a.
$$\sum_{dihedrals} [V_1(1+\cos(\varphi))+V_2(1-\cos(2\varphi))+V_3(1+\cos(3\varphi))] +$$	→ b.
$$\sum_{atom\,i}\sum_{atom\,j}\left\{q_iq_j/r_{ij}+4\varepsilon_{ij}\left[(\sigma_{ij}/r_{ij})^{12}-(\sigma_{ij}/r_{ij})^6\right]\right\}$$	→ c.

a. Molecular geometry: bond lengths and angles

The molecular geometry parameters are generally obtained from low-level *ab-initio* calculations (HF, B3LYP) using a simple basis set (6-31G) describing a single molecule. All the models under discussion used a similar method to estimate the parameters describing the imidazolium ring and the published results are comparable with experimental data obtained by diffraction studies as shown in Table III. Two conclusions can be drawn directly from the table: i) the ring geometry is not strongly affected by the environment of the imidazolium molecule since the ring geometry in two completely different crystals, [emim][VOCl₄] and [ddmim][PF₆], (10,11) are comparable between them and similar to the *ab-initio* values for the isolated imidazolium cation; and ii) the distortion of the ring caused by different alkyl substituents is so small that the use of a symmetrical ring geometry as in a mmim$^+$ cation represents a good approximation.

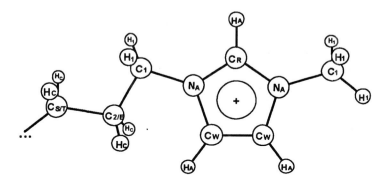

Figure 1. Nomenclature of the 1-alkyl-3-methyl-imidazolium cations.In the side chains a=alkyl, m=methyl, e=ethyl, b=butyl, h=hexyl, dd=dodecyl.

Most bond and angle parameters used in this work were taken from or based on the OPLS-AA and AMBER force fields (7–9,16), which are largely compatible with respect to the intramolecular terms. Whenever we found significant differences between our *ab-initio* geometries and OPLS-AA or AMBER parameters, notably in equilibrium distances and angles, we proposed new values, reported in Table IV. We did not modify force constants from what is specified in AMBER and OPLS-AA, since these features have a relatively small effect on thermodynamic properties. The main departure from the OPLS-AA force field at this point regards our use of constrained bond lengths for all C–H stretching modes present in the cations.

Margulis *et al.* (6) also used OPLS-AA/AMBER parameters to define the internal force field of the 1-butyl-3-methyl imidazolium cation but chose to model the ring directly with imidazole parameters (neutral and asymmetric ring). The same kind of remark applies to the work of de Andrade *et al.(3)*: although the CHARMM forcefield contains parametric data concerning the protonated histidine ring, the authors chose parameters that can be understood as averages of the two (asymmetrical) halves of the neutral histidine ring. Unlike the imidazolium ring, the imidazole ring residue is not symmetrical and the distortions of the structure around the carbon connected to the two (non-equivalent) nitrogen atoms are noticeable. The use of imidazole bond distances and angles to define an imidazolium cation can lead to an inaccurate parameterization.

To establish the molecular geometry of the PF_6^- and NO_3^- anions (taken as octahedral and planar triangular structures, respectively) it is sufficient to know the P–F and N–O interatomic distances. These were taken from *ab-initio* data (*13,14*) and are also presented in Table III. The corresponding distances obtained from x-ray diffraction studies in selected IL crystals compare favorably with the *ab-initio* results.

Table III. Comparison between experimental x-ray (XR) geometries and single-molecule *ab-initio* (AI) calculations

	$emim^{+\,a1}$ XR (10)	$ddmim^{+a2}$ XR (11)	$emim^+$ AI (3)	$emim^+$ AI (4)	$amim^+$ AI (7)
bonds (Å)[b]					
N_A–C_R	1.311(4)	1.322(3)	–	1.314	**1.315**
	1.311(4)	1.326(3)		1.315	
N_A–C_W	1.357(5)	1.373(3)	–	1.378	**1.378**
	1.360(6)	1.374(3)		1.378	
C_W–C_W	1.334(8)	1.334(8)	–	1.342	**1.341**
N_A–C_1	1.452(4)	1.468(3)	–	1.466	**1.466**
	1.468(4)	1.477(3)		1.478	
C_1–C_E	1.500(9)	–	–	1.520	–
angles (deg)[b]					
N_A–C_R–N_A	109.6(3)	–	109.8	109.9	**109.8**
N_A–C_W–C_W	107.1(3)	–	108.1	107.0	**107.1**
	107.6(4)		108.0	107.2	
C_W–N_A–C_R	108.0(3)	–	106.9	107.9	**108.0**
	107.6(2)		106.8	108.0	
C_W–N_A–C_1	125.9(3)	–	–	125.6	**125.6**
	125.2(3)			125.9	
C_R–N_A–C_1	126.5(3)	–	–	126.4	**126.3**
	125.4(3)			126.1	
	PF_6^{-c1} XR (12)	PF_6^{-a2} XR (11)	PF_6^- AI (13)	NO_3^{-c2} XR (14)	NO_3^- AI (15)
bonds (Å)					
P–F	1.566	1.596	**1.560**	–	–
N–O	–	–	–	1.220	**1.250**

[a1] [emim][VOCL$_4$] crystal ; [a2] [ddmim][PF$_6$] crystal.

[b] double entries refer to bond lengths and angles on the alkyl and methyl side of the imidazoilum ring, respectively.

[c1] [emim][PF$_6$] crystal ; [c2] [emim][NO$_3$] crystal.

Of course one might object that since the cations were modeled as flexible molecules, we could have used a similar approach when dealing with the anions. Two facts justify our present choice: i) the anions under discussion are not articulated molecules. This means that using fixed bond lengths and angles will not hinder any internal movement of the molecule; and ii) there is a large uncertainty associated with the parameterization of the external force field of these ions. In our opinion the use of a more complete internal force field is not relevant at this stage, before these (and other) anions have a better defined set of external parameters.

b. *Ab-initio* calculations of torsion energy profiles

The parameters defining the torsion energy profiles in the imidazolium ring were taken from the AMBER force field concerning heterocyclic aromatic

Table IV. Stretching and bending force-field parameters

Bonds	r_{eq} (Å)	K_r $(kJ\,mol^{-1}\text{Å}^{-2})$	
$C_{R/W}-H_A$	1.08	-	OPLS imidazole (16), constrained
C^*-H^* [a]	1.09	-	OPLS butane (8), constrained
C_R-N_A	1.315	1996	This work, OPLS imidazole (16)
C_W-N_A	1.378	1787	This work, OPLS imidazole (16)
C_W-C_W	1.341	2176	This work, OPLS imidazole (16)
N_A-C_1	1.466	1410	This work, AMBER N*-CT (9,16)
C^*-C^* [a]	1.529	1121	OPLS butane (8)
Angles	θ_{eq} (deg)	$K_\theta (kJ\,mol^{-1}rad^{-2})$	
$C_W-N_A-C_R$	108.0	292.6	This work, OPLS imidazole (16)
$C_W-N_A-C_1$	125.6	292.6	This work, OPLS imidazole (16)
$C_R-N_A-C_1$	126.4	292.6	This work, OPLS imidazole (16)
$N_A-C_R-H_A$	125.1	146.3	This work, OPLS imidazole (16)
$N_A-C_R-N_A$	109.8	292.6	This work, OPLS imidazole (16)
$N_A-C_W-C_W$	107.1	292.6	This work, OPLS imidazole (16)
$N_A-C_W-H_A$	122.0	146.3	This work, OPLS imidazole (16)
$C_W-C_W-H_A$	130.9	146.3	This work, OPLS imidazole (16)
$N_A/C^*-C^*-H^*$ [a]	110.7	313.2	OPLS butane (8)
$N_A/C^*-C^*-C^*$ [a]	112.7	418.4	OPLS butane (8)
$H^*-C^*-H^*$	107.8	276.1	OPLS butane (8)

[a] no difference found in the equilibrium angles and force constants of aliphatic carbons connected either to carbon or nitrogen. C^* represents a generic aliphatic carbon, either C_1, C_2, C_E, C_S or C_T. H^* represents either H_1 or H_C.

molecules. These include the improper dihedrals needed to maintain the ring and the atoms directly attached to it in a planar geometry. The OPLS-AA specification of five-membered heterocyclic aromatic rings (*16*) constrained the ring and the atoms directly connected to it to be planar, which means that no dihedrals involving those atoms needed to be defined. Nevertheless, the authors of OPLS-AA recommend the use of AMBER parameters for cases where there are no ad-hoc constraints. The torsion profiles along the aliphatic side chain were obtained from the published OPLS-AA results (*8*). The missing parameters corresponding to the dihedrals between atoms belonging to the imidazolium ring and those in the alkyl side chain were defined in this work with the aid of ab-intio calculation.

Table V. Torsional force-field parameters

Dihedrals	V_1 $(kJmol^{-1})$	V_2 $(kJmol^{-1})$	V_3 $(kJmol^{-1})$	
$X-N_A-C_R-X$	0	19.46	0	AMBER (9)
$X-C_W-C_W-X$	0	44.98	0	AMBER X–CC–CW–X (9)
$X-N_A-C_W-X$	0	12.55	0	AMBER (9)
$C_W-N_A-C_1-H_1$	0	0	0.55	this work
$C_R-N_A-C_1-H_1$	0	0	0	this work
$C_W-N_A-C_1-C_{2/E}$	−5.76	4.43	0.877	this work
$C_R-N_A-C_1-C_{2/E}$	−3.23	0	0	this work
$N_A-C_1-C_2-C_{S/T}$	0.738	−0.681	1.02	this work
$N_A-C_1-C_{2/E}-H_C$	0	0	0	this work
$C^*-C^*-C^*-H^*$	0	0	1.531	OPLS butane (8)
$H^*-C^*-C^*-H^*$	0	0	1.331	OPLS butane (8)
$C^*-C^*-C^*-C^*$	0.728	−0.657	1.167	OPLS butane (8)
$X-N_A-X-X$ [b]	0	8.37	0	AMBER X–N*–X–X (9)
$X-C_{W/R}-X-X$ [b]	0	9.2	0	AMBER (9)

[a] C^* represents a generic aliphatic carbon, either C_1, C_2, C_E, C_S or C_T. H^* represents either H_1 or H_C.

[b] improper dihedral.

New torsion potentials corresponding to dihedral angles between the imidazolium ring and the alkyl side chains were developed following a procedure similar to the one adopted for the OPLS-AA force field (*8,17*). They were obtained from relaxed potential energy scans at a series of values for the dihedral angles of interest, that were held fixed while the remaining internal coordinates of the molecule were allowed to relax to energy minima. Quantum chemical calculations were performed using Gaussian 98 (*18*) at the MP2/cc-pVTZ(-f)//HF/6-31G(d) level of theory, thus using the same basis set as in the recent OPLS-AA model for perfluoroalkanes (*17*), and are documented in detail elsewhere (*7*).

The parameters obtained in the present study are collected in Table V The present numerical coefficients were calculated specifically for the imidazolium cation family, using the building-up procedure explained in previous publications (7,21), whenever parameters were not available in OPLS-AA or AMBER.

The task of fitting force field parameters to torsion energies is particular and specific, in the sense that the non bonded interactions (Lennard-Jones and coulombic) between atoms separated by three bonds comes into play and participate with a share of the rotation barriers. The significance of this non-bonded contributions is not easily estimated a priori (21). Therefore, contrary to the cases of bond lengths and valence angles for which parameters can be taken from different sources without major concerns, the torsion energy profiles depend on the cosine series (cf. Table II) but also on the intermolecular features of the model. It is not correct to transpose the parameters of the cosine series between different models: these should be readjusted to attain agreement with the non-bonded part.

In conclusion to this section, we justify the effort undertaken here, for the first time to our knowledge, to calculate in a rigorous manner the dihedral terms affecting the junction between the aromatic ring and the alkyl side chain for the family of imidazolium cations. The force fields proposed in the literature so far contain some crude approximations and guesswork in this particular aspect, that certainly affect the conformational details of the ionic liquids.

Table VI External force field parameterization

Authors (ref.)	HPLB (2)	ABS (3)	SBM (4)	MM (5)	MSB (6)	CLDP (7)
External force field						
Functional	Exp6+q	LJ+q	LJ+q	LJ+q	LJ+q	LJ+q
Cations	Williams	AMBER	OPLS-UA	CHARMM	OPLS-AA	OPLS-AA
Anions	Williams, MMFF94	AMBER, DREIDING	SF_6 data	CHARMM	OPLS-AA	OPLS-AA, AMBER
Electron density						
Theory level	MP2	UHF	RHF	B3LYP	HF	MP2
Basis set	6-31G**	6-31G*	6-31G*	6-311+G*	6-31G**	cc-pVTZ
Charges	DMA	RESP	CHelpG	CHelpG	ESP	CHelpG

[a] The five atoms in the ring plus the five atoms directly attached to them.

c. Intermolecular potential

The OPLS-AA / AMBER external force fields both comprise repulsion-dispersion terms parameterized by a 12-6 Lennard Jones potential function and

Table VII. 1-Alkyl-3-methylimidazolum non.bonded interactions parameters

Atoms	Q (e)	$\sigma(\text{Å})$	ε (kJmol^{-1})	Source
Cations				
C_1	−0.17	3.50	0.2761	This work, OPLS butane (8)
C_2	0.01	3.50	0.2761	this work, OPLS butane (8)
C_E	−0.05	3.50	0.2761	this work, OPLS butane (8)
C_R	−0.11	3.55	0.2929	this work, OPLS HIP^a(17)
C_S	−0.12	3.50	0.2761	OPLS butane (8)
C_T	−0.18	3.50	0.2761	OPLS butane (8)
C_W	−0.13	3.55	0.2929	this work, OPLS HIP^a(17)
H_A	0.21	2.42	0.1255	this work, OPLS HIP^a(17)
H_C	0.06	2.50	0.1255	OPLS butane (8)
H_1	0.13	2.50	0.1255	this work, OPLS butane (8)
N_A	0.15	3.25	0.7113	this work, OPLS HIP^a(17)
Anions				
P, PF$_6^-$	1.34	3.74	0.8368	OPLS (8)
F, PF$_6^-$	−0.39	3.12	0.2552	OPLS (8)
N, NO$_3^-$	0.95	3.06	0.3380	Fit to Born-Mayer+q_{A1} (24)
O, NO$_3^-$	−0.65	2.77	0.6100	Fit to Born-Mayer+q_{A1} (24)
Cl, Cl$^-$	−1.00	3.77	0.6200	Fit to Born-Mayer (25)

^{a} HIP is the protonated histidine cation.

an electrostatic term represented by partial charges located at each interaction site of the molecule. This intermolecular potential functional was employed by all but one of the models under review, cf Table VI.

In the case of the imidazolium cations, the Lennard-Jones parameters for each type of atom were taken from the OPLS-AA parameterization of heterocyclic aromatic rings (16) or aliphatic compounds (8) (Table VII). The interactions between atoms of different type were parameterized using the Lorentz-Berthelot mixing rules (arithmetic and geometric mean rules for σ and ε, respectively). The partial charges on each atom were calculated from the electron density obtained by *ab-initio* calculation, using an electrostatic surface potential methodology (7).

Table VIII. Charge distribution in IL cations according to different force fields. All numbers are percentual values of the charge of one proton

Authors (ref.)	HPLB (2)	ABS (3)	SBM [a] (4)	MM (5)	MSB (6)	CLDP (7)
Cations	$mmim^+$, $emim^+$	$emim^+$, $bmim^+$	$bmim^+$	$bmim^+$	$bmim^+$	$amim^+$ (C_1-C_{12})
Ring atoms						
N_A	-27	4	10	12	15	15
C_R	41	3		6	-6	-11
H_A	10	22	23	18	24	21
C_W	11	-18		-18	-20	-13
H_A	9	25	6	19	24	21
Side-chain atoms						
C_1	12	-9		-4	-25	-17
H_1	6	11	22	8	16	13
C_E	-6	-8		-	-	-5
H_C	5	6	-	-	-	6
C_2	-	3		-12	no data	1
H_C	-	2	12	3	no data	6
C_S	-	5		26	no data	-12
H_C	-	2	12	-6	no data	6
C_T	-	-16		-21	0	-18
H_C	-	5	-5	6	4	6

[a] United atom potential

The partial charges and Lennard-Jones parameters in the PF_6^- anion were taken directly from the OPLS-AA forcefield (8). The NO_3^- ion was modeled by four Lennard-Jones centers whose parameters were determined by fitting to the Born-Mayer potential used to describe dispersive interactions in molten salts (18). The partial charges had been calculated *ab initio*. The parameters for Cl⁻ were obtained in a similar way, fitting to the Born-Mayer potentials developed for crystalline and molten salts (7, 25). At this point it is important to stress that the resulting Lennard-Jones parameters for Cl⁻ are quite different from those usually employed for chlorine atoms bonded to organic molecules or aqueous Cl⁻. The differences are apparent if we compare the interaction diameter, σ, in the three situations: 3.77 Å in molten salts / ionic crystals (25), around 3.5 Å in organic chlorine atoms (2,8,9) and around 4.4 Å in aqueous solution (9).

Ab-initio calculation of partial charges

Atomic charges on the imidazolium ring and its adjacent atoms were also obtained in the present study. They were calculated for mmim$^+$, emim$^+$ and bmim$^+$ by electrostatic surface potential fits, using the CHelpG procedure (*26*), to electron densities obtained at the MP2/cc-pVTZ(-f) level (*7*). In order to ensure the transferability of the model along the entire dialkylimidazolium cation family, i) the charges on the atoms of the ring, those directly attached to it and the H_l hydrogen atoms, were given symmetrical values along the C_{2v} axis of the ring and have values close to those found for mmim$^+$. For bmim$^+$ and emim$^+$ the differences, including departures from symmetry, are small; ii) the charges on all atoms in the alkyl side chain removed from the ring more than 3 bonds (C_S or C_T and H_C) were given the corresponding OPLS-AA values for alkanes (*8*); iii) the charge in the remaining atom that establishes the connection between the ring and the side chain, C_E in emim$^+$ and C_2 in longer side chains, was found empirically in order to respect the total charge of the cation. The values of the adopted charges and comparisons with other force fields are shown in Tables VII and VIII..

Although we performed calculations on a larger basis set and employed a method including electron correlation (MP2), the charges obtained in this work show an overall agreement with those proposed by several other authors that used electrostatic surface potential fits to the *ab-initio* electron densities. An exception is the relative charge distribution between the nitrogen atoms and carbon atom between them: our N_A atoms are somewhat more positive and this is compensated by a less positive C_R atom. The distributed multipole analysis method employed by Hanke *et al.* (*2*) yielded a significantly different set of atomic charges, both in sign and magnitude, when compared with all other authors.

Validation

MD Simulations

The validation of the proposed force field was based on simulation results obtained using the molecular dynamics technique, implemented with the DL_POLY software (*23*). The strategy was simple: since the most accurate and common experimental data concerning ILs are densities, we decided to test the performance of the proposed force field in the estimation of molar densities along the family of the imidazolium cation, with different counter ions, both in the crystalline and liquid phases.

Crystaline phase

The ionic liquids based on the 1-alkyl-3-methylimidazolium cation family exhibit higher melting-point temperatures for cations with either very short or long alkyl side-chain lengths. The crystalline structures of those salts can be used to validate the forcefield parameterization. Such procedure was used by Hanke *et al.* (2) to test their model describing the above mentioned mmim$^+$ and emim$^+$ salts. In the present work five crystalline structures containing the imidazolium cation were selected from the Cambridge Crystallographic Database (CCDB). The simulation boxes and initial configurations were set taking into account the dimensions and occupancy of the unit cells of each crystalline structure (7). The simulation details and results for each crystal are summarized in Table IX.

Table IX. Crystalline phase simulation results

Salt	*[mmim]* Cl	*[emim]* Cl	*[emim]* [NO$_3$]	*[emim]* [PF$_6$]	*[ddmim]* [PF$_6$]
crystallographic references (from CCDB)					
Ref	(27)	(28)	(14)	(12)	(11)
Sp. Group	14b	19	14b	14c	14a
Simulation details					
# pairs	192	144	96	192	72
# cells	4 x 4 x 3	3 x 3 x 1	6 x 2 x 2	4 x 4 x 3	3 x 3 x 2
cutoff (Å)	13.5	13.5	12.5	16	13.5
Temp.(K)	203	298	298	298	123
Crystallographic versus simulated data					
a (Å)	8.652 **8.95**	10.087 **10.13**	4.540 **4.56**	8.757 **8.82**	9.175 **9.10**
b (Å)	7.858 **7.75**	11.179 **11.07**	14.820 **14.1**	9.343 **9.20**	9.849 **9.75**
c (Å)	10.539 **10.3**	28.773 **28.8**	13.445 **13.4**	13.701 **13.6**	22.197 **23.6**
β (deg)	106.34 **108.9**	90 **90.0**	95.74 **93.1**	103.05 **102.3**	94.132 **93.0**
V (Å^3)	687.58 **698**	3240.0 **3228**	899.4 **860**	1092.0 **1083**	2000.6 **2081**
ρ (mol/dm^3)	9.660 **9.49 0.02**	8.200 **8.23 0.03**	7.385 **7.73 0.05**	6.083 **6.13 0.03**	3.320 **3.19 0.07**
δρ (%)	−1.8	+0.4	+4.7	+0.8	−3.9
U (kJ/mol)	−438.8 0.5	−431.4 0.9	−412.0 1.2	−367.8 0.6	−402 2

The objective of these simulations is not to test the ability of the present force field to generate the corresponding experimental crystal lattice — a procedure that is still controversial even for simple molecular crystals — but simply to check if they are compatible. The stringency of the test was confirmed with simulation runs where ad-hoc parameters were introduced and large distortions of the unit cell paramenters of the lattice observed. The use of an anisotropic barostat coupled with the system allows for a relatively short equilibration period due to the frequent rescaling of the position of the particles. The rather short but effective relaxation time was also confirmed by monitoring the length of the sides and angles of the simulation box as the simulation proceeded through the equilibration period.

Table X. Liquid phase simulation results

Salt	[emim] [PF$_6$]	[emim] [NO$_3$]	[emim] Cl	[bmim] [PF$_6$]	[bmim] [NO$_3$]
Exp. method (ref.)	Unknown (30)	–	–	Picnometry (29)	Picnometry (31)
ρ (mol/dm³)	6.013ᵃ 5.67±0.01	– 7.32±0.02	– 7.67±0.02	4.796 4.71±0.01	5.71±0.04 5.78±0.01ᵇ
δρ (%)	–5.7ᵃ	–	–	–1.9	+1.2
U (kJ/mol)	–352.9±0.1	–401.3±0.4	–403.8±0.3	–335.9±0.8	–380.7±0.3

Salt	[bmim] Cl	[hmim] [PF$_6$]	[hmim] [NO$_3$]	[hmim] Cl	
Exp. method (ref.)	Gravimetry (30)	Picnometry (29)	–	Gravimetry (30)	
ρ (mol/dm³)	6.183 6.00±0.02	4.186 4.05±0.01	– 4.85±0.01	5.081 4.94±0.02	
δρ (%)	–3.0	–3.0	–	–3.0	
U (kJ/mol)	–387.3±0.5	–359.1±0.6	–405.5±0.4	–410.0±0.6	

ᵃ Experimental liquid density at unknown temperature. Probably above the reported melting point temperature of 331–333 K.

ᵇ Experimental and simulation results obtained at 313 K.

Liquid Phase.

The ionic liquids simulated were the combinations of $emim^+$, $bmim^+$ and $hmim^+$ with PF_6^-, NO_3^- and Cl^-. Technical details are given elsewhere (*7*). Simulation and relevant experimental data are presented in Table X.

Discussion

The density results shown in Tables IX and X exhibit relative deviations from the corresponding experimental values of the order of a few units of percent (1–5 %). These deviations are of the same order of magnitude as those obtained by other authors when comparing the performance of their models against experimental density data (*2–6*). In our case we used the same framework to calculate the density of fourteen different ionic liquids — of which nine could be compared against experimental data — both in the liquid and crystalline phases. The results are consistent along the imidazolium cation family ($mmim^+$, $emim^+$, $bmim^+$, $hmim^+$ and $ddmim^+$) with three different anions. The level of agreement is very good considering that the calculations are purely predictive: all parameters used were either taken as such from the OPLS-AA/AMBER force field or calculated *ab initio*, none was adjusted to match experimental data.

The structural properties of the crystals considered in the present study were also correctly predicted by the model. After relaxation, all crystallographic parameters were reproduced within uncertainties corresponding to a few tenths of Å (except for $ddmim^+$ where deviations are somewhat larger) in the length of the sides of the unit cells and up to two degrees in the β director angle of the monoclinic crystals.

At this point we think that the discussion of other results obtainable by simulation with the present model — structural data obtained *via* the calculation of radial distribution functions, density as a function of temperature or pressure, ionic self-diffusion coefficients or even the internal energy of the system — is unwarranted due to the scarcity or inexistence of reliable and direct experimental results. This means that while the volumetric behavior of the ILs of the imidazolium family is correctly captured by our model, the properties closely related to energetic characteristics are difficult to validate at present. The former are connected to structural characteristics of the present force field that seem to be soundly established, while the latter depend significantly on the attribution of electrostatic interaction sites where a complete answer has yet to be formulated. In the present effort we sought to contribute to the discussion of this problem by providing a complete set of atomic partial charges from *ab-initio* calculations including electron correlation on an extended basis set.

As such, the present model can be regarded as a step towards a general force field for ILs of the imidazolium family that was built in a coherent way, is easily integrated with OPLS-AA/AMBER, is transferable between different combinations of cation/anion and was validated against available solid and liquid state properties. The extension of the present force field to other IL families, namely imidazolium cations substituted with non-alkyl side chains, would require the calculation of new intramolecular and charge distribution parameters using a *ab initio* / MD methodoly similar to the employed in this work.

References

1. (a) Welton, T. *Chem. Rev.* **1999**, *99*, 2071; (b) Holbrey, J. D.; Seddon, K. R. *Clean Prod. Process.* **1999**, *1*, 223; (c) Brennecke, J. F.; Maginn, E. J. *AIChE J.* **2001**, *47*, 2384; (d) Huddleston, J. G.; Visser, A. E.; Reichert, W. M.; Willauer, H. D.; Broker, G. A.; Rogers, R. D. *Green Chem.* **2001**, *3*, 156; (e) Seddon, K. R.; Stark, A.; Torres, M.-J. *Pure Appl. Chem.* **2000**, *72*, 2275
2. Hanke, C. G.; Price S. L.; Lynden-Bell, R. M. *Mol. Phys.* **2001**, *99*, 801.
3. (a) de Andrade, J.; Böes, E. S.; Stassen, H. *J. Phys. Chem. B* **2002**, *106*, 3546; (b) de Andrade, J.; Böes, E. S.; Stassen, H. *J. Phys. Chem. B* **2002**, *106*, 13344.
4. Shah, J. K.; Brennecke, J. F.; Maginn, E. J. *Green Chemistry* **2002**, *4*, 112.
5. Morrow T. I.; Maginn E. J. *J. Phys. Chem. B* **2002**, *106*, 12807.
6. Margulis, C. J.; Stern H. A.; Berne B. J. *J. Phys. Chem. B* **2002**, *106*, 12017.
7. Canongia Lopes, J. N.; Deschamps, J.; Pádua, A. A. H. ,accepted for publication in *J. Phys. Chem. B*.
8. Jorgensen, W. L.; Maxwell, D. S.; Tirado-Rives, J. *J. Am. Chem. Soc.* **1996**, *118*, 11225 ; Kaminski, G.; Jorgensen, W. L. *J. Phys. Chem* **1996**, *100*, 18010.
9. (a) Cornell, W. D.; Cieplak, P.; Bayly, C. I.; Gould, I. R.; Merz, K. M.; Ferguson, D. M.; Spellmeyer, D. C.; Fox, T.; Caldwell, J. W.; Kollman, P.A. *J. Am. Chem. Soc.* **1995**, *117*, 5179. (b) Parameters obtained from file parm99.dat corresponding to AMBER versions 1999 and 2002.
10. Hitchcock, P. B.; Lewis, R. J.; Welton, T. *Polyhedron* **1993**, *12*, 2039.
11. Gordon, C. M.; Holbrey, J. D.; Kennedy, A. R.; Seddon, K. R. *J. Mater. Chem.* **1998**, *8*, 2627.
12. Fuller, J.; Carlin, R. T.; De Long, H. C.; Haworth, D. *J. Chem. Soc. Chem. Commun.* **1994**, 299.
13. Kaminski G.; Jorgensen, W. L. *J. Chem. Soc. Perkin Trans 2* **1999**, 2365.
14. Wilkes, J. S.; Zaworotko, M. J. *J. Chem. Soc. Chem. Commun.* **1992**, 965.
15. Vöhringer, G.; Richter, J. Z. *Naturforsch.* **2001**, *56a*, 337.
16. McDonald, N. A.; Jorgensen, W. L. *J. Phys. Chem. B* **1998**, *102*, 8049.
17. (a) Watkins, E. K.; Jorgensen, W. L. *J. Phys. Chem. A* **2001**, *105*, 4118; (b) Rizzo R. C.; Jorgensen W. L., *J. Am. Chem. Soc.* **1999**, *121*, 4827.

18. Frisch, M. J.; Trucks, G. W.; Schlegel, H. B.; Scuseria, G. E.; Robb, M. A.; Cheeseman, J. R.; Zakrzewski, V. G.; Montgomery, Jr., J. A.; Stratmann, R. E.; Burant, J. C.; Dapprich, S.; Millam, J. M.; Daniels, A. D.; Kudin, K. N.; Strain, M. C.; Farkas, O.; Tomasi, J.; Barone, V.; Cossi, M.; Cammi, R.; Mennucci, B.; Pomelli, C.; Adamo, C.; Clifford, S.; Ochterski, J.; Petersson, G. A.; Ayala, P. Y.; Cui, Q.; Morokuma, K.; Salvador, P.; Dannenberg, J. J.; Malick, D. K.; Rabuck, A. D.; Raghavachari, K.; Foresman, J. B.; Cioslowski, J.; Ortiz, J. V.; Baboul, A. G.; Stefanov, B. B.; Liu, G.; Liashenko, A.; Piskorz, P.; Komaromi, I.; Gomperts, R.; Martin, R. L.; Fox, D. J.; Keith, T.; Al-Laham, M. A.; Peng, C. Y.; Nanayakkara, A.; Challacombe, M.; Gill, P. M. W.; Johnson, B.; Chen, W.; Wong, M. W.; Andres, J. L.; Gonzalez, C.; Head-Gordon, M.; Replogle, E. S.; Pople, J. A. *Gaussian 98 (Revision A.10)*, Gaussian, Inc., Pittsburgh PA, **2001**.

19. Dunning, T. H. Jr. *J. Chem. Phys.* **1989**, *20*, 1007.

20. Wang, J.; Cieplak, P., Kollman, P. A. *J. Comp. Chem.* **2000**, *21*, 1099.

21. Pádua, A. A. H. *J. Phys. Chem. A* **2002**, *106*, 10116.

22. Friesner, R. A.; Murphy, R. B.; Beachy, M. D.; Ringnalda, M. N.; Pollard, W. T.; Drunietz, B. D.; Cao, Y. *J. Phys. Chem. A* **1999**, *103*, 1913.

23. Smith, W.; Forester, T. R. *The DL_POLY package of molecular simulation routines, version 2.12*, The Council for The Central Laboratory of Research Councils, Daresbury Laboratory, Warrington, UK, **1999**.

24. Signorini, G. F.; Barrat, J.-L.; Klein, M. *J. Chem. Phys.* **1990**, *92*, 1294.

25. Tosi, M. P.; Fumi, G. *J. Phys. Chem. Solids* **1964**, *25*, 45.

26. Brenneman, C. M.; Wiberg, J. *J. Comp. Chem.* **1987**, *8*, 894.

27. Arduengo III, A. J.; Dias, H. V. R.; Harlow, R. L.; Kline, M. *J. Am. Chem. Soc.* **1992**, *114*, 5530.

28. Dymeck Jr, C. J.; Grossie, D. A.; Fretini, A. V.; Adams, W. W. *J. Mol. Struct.* **1989**, *213*, 25.

29. Chung, S.; Dzyuba, S. V.; Bartsch, R. A. *Anal. Chem.* **2001**, *73*, 3737.

30. Huddleston, J. G.; Visser, A. E.; Reichert, W.; Wilauer, M.; Heather, D.; Brocker, G. A.; Rogers, R. D. *Green Chemistry* **2001**, *3*, 156.

31. Blanchard, L. A.; Gu, Z.; Brennecke, J. F., *J. Phys. Chem B* **2001**, *105*, 2437.

Chapter 11

Interactions of Gases with Ionic Liquids: Molecular Simulation

Johnny Deschamps and Agilio A. H. Pádua

Laboratoire de Thermodynamique des Solutions et des Polymères,
Université Blaise Pascal, Clermont-Ferrand, 63177 Aubière, France

The interactions of argon, methane, nitrogen, oxygen and
carbon dioxide with room temperature ionic liquids of the
dilakylimidazolium family, 1-butyl-3-methylimidazloium
hexafluorophosphate, [bmim][PF$_6$], and 1-butyl-3-methyl-
imidazolium tetrafluoroborate, [bmim][BF$_4$], were studied by
molecular simulation. The ionic liquids were modeled using a
recently proposed force field and the solute gases represented
by multi-site models from the literature. The calculated
solubilities are in qualitative agreement with experiment
concerning the relative magnitude of this property for the
different gases. The temperature dependence of the solubilities
is correctly predicted for carbon dioxide but has the wrong
sign for the remaining gases. Particular attention is devoted to
the structure of the solutions of carbon dioxide in the ionic
liquids. It was observed that the solubility of carbon dioxide
can be correctly predicted from physical interaction only. Our
results show that the molecule of carbon dioxide does not
interact in a relevant manner with the hydrogen connected to
the C$_2$ carbon of the imidazolium ring (the carbon atom
bonded to the two nitrogens), as could be expected.

The solvation properties of room temperature ionic liquids are being elucidated by a number of experimental and theoretical studies. The present work concerns the interactions of gases with ionic liquids and was closely articulated with an experimental study (*1*). Here those interactions are investigated using atomistic computer simulation, with two objectives: (i) to verify the performance of a force field newly proposed for imidazolium salts in predicting properties that depend on unlike interactions, and (ii) to contribute to understand the nature of solute-solvent interactions.

Computational Methods

The ionic liquids 1-butyl-3-methylimidazloium hexafluorophosphate, [bmim][PF$_6$], and 1-butyl-3-methylimidazolium tetrafluoroborate, [bmim][BF$_4$], were modeled by an all-atom force field developed recently for the family of dialkylimidazolium salts (*2*). The force field is derived from OPLS-AA (*3*) and AMBER (*4*): it adopts the same functional form and, whenever available, parameters were taken from these two sources. The solute gases were represented by intermolecular potentials of the type Lennard-Jones plus charges, obtained from the literature. Argon was represented by a Lennard-Jones potential (*5*), nitrogen and oxygen were modeled by two centers (*6,7*), methane was considered as defined in OPLS-AA (*3*), and for carbon dioxide the rigid three-site model with partial charges proposed by Harris and Yung was adopted (*8*). No adjustments of unlike interaction parameters were performed within this work, therefore all results presented below are predictive.

Simulations were performed using the molecular dynamics method, implemented in the DL_POLY package (*9*). Solubilities were obtained from the residual chemical potentials of the solute gases at infinite dilution, calculated using a free energy route such as the test-particle insertion method. The chemical potential was calculated at infinite dilution, in systems constituted by 250 ion pairs. Simulations were run for 350 ps, in the *NpT* ensemble at 1 bar, and samples of 1000 configurations were stored for post-treatment. In each configuration, 90 000 insertions at random positions were attempted. Radial distribution functions were obtained from a different set of runs lasting 200 ps, of systems containing one solute molecule and 250 ion pairs. Previous to the production runs, the systems were allowed to equilibrate for at least 200 ps. A time step of 1 fs was adopted and a spherical cutoff of 13 Å for atomic interactions was considered. Tail corrections to short-range interactions were included and the long-range coulombic interactions were calculated by the Ewald summation method. Except for C–H bond lengths, which were constrained, all other internal modes in the imidazolium cations were left flexible.

Results and Discussion

Solubility of gases

The solubility of the five gases calculated in the ionic liquids [bmim][PF$_6$] and [bmim][BF$_4$], at 323 K, is presented in Figure 1 and compared to experimental data obtained by Brennecke and co-workers in [bmim][PF$_6$] (*10*). Although agreement is not quantitative, the relative solubility order for the different gases is reproduced by simulation, with CO$_2$ showing a much larger solubility, and N$_2$ a lower solubility, than Ar, CH$_4$ and O$_2$. Experimental values for N$_2$ were not published because its solubility is below the detection limit of the measuring device (*10*).

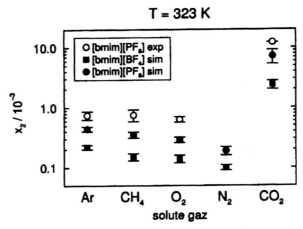

Figure 1. Comparison of calculated gas solubilities in the ionic liquids with experimental values. Solubility is given as the mole fraction of gas dissolved at a partial pressure of 1 bar.

The present results indicate that the solubility of gases is higher in [bmim][PF$_6$] than in [bmim][BF$_4$]. Spectroscopic evidence for CO$_2$ (*11*), however, shows that its interaction is stronger with BF$_4^-$ than with PF$_6^-$, due to the smaller size of the former anion. The same authors postulate that this effect might be compensated by free-volume contributions, which would be larger in [bmim][PF$_6$] (*11*), explaining thus the higher solubility of CO_2 in imidazolium ionic liquids with the PF$_6^-$ anion when compared to those with BF$_4^-$ anion (*12*).

We have attempted in this work to address the issue of free volume in the ionic liquids, through the calculation of free energy of cavity formation. The solvation process is commonly decomposed in two steps: creation of a cavity in the solvent capable of hosting a solute molecule, and activation of solute-solvent interactions. The free energy of cavity formation can be obtained from simulations of the pure solvent, using the technique of test particle insertion to find the probability of inserting a hard sphere at random locations (*13*). The free energy of cavity formation in the two ionic liquids calculated in this work is shown in Figure 2, where it is compared to that of an organic solvent, *n*-hexane, and water. These two liquids were represented by OPLS-AA (*3*) and TIP5P (*14*) models, respectively.

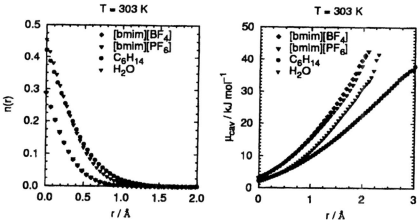

Figure 2. Distribution of spherical cavity sizes (left) and free energy of cavity formation(right)as a function of cavity size in the ionic liquids compared to that in other solvents.

It is observed that spontaneously present cavities are of a smaller size in the ionic liquids than in *n*-hexane or water at the same temperature. Accordingly, reversible creation of a cavity requires more work in the ionic liquids than in the two other fluids. As a result, if solute-solvent interactions were similar, any gas would be less soluble in the ionic liquids than in the two other solvents. Since CO_2 has a larger molecular size than any of the other gases considered in this work, it is clear that it is the strength of its interactions with the ionic liquids that is the cause of its higher solubility.

Both ionic liquids considered in this study have very similar distributions of cavity sizes, meaning that the free volume available in the two should be comparable. Nevertheless, from the plot of free energy of cavity formation it appears indeed to be slightly easier to create cavities in [bmim][PF$_6$] than in [bmim][BF$_4$]. This agrees with the view expressed in the literature (11,12).

In Figure 3 are plotted the calculated solubilities of Ar and CO$_2$ in the two ionic liquids, as a function of temperature, compared to experimental results (1). It is observed that the temperature dependence of the solubility of Ar obtained from simulation is opposite to the experimental trend. For CO$_2$, simulation agrees with experiment, both giving solubilities that decrease with temperature. The temperature dependence of the solubility is related to the enthalpy of solvation, therefore the interaction models used do not yield the correct sign for the enthalpy of solvation of Ar. We have yet no explanation for this fact, since it appears to indicate that, in our simulations, the interactions of the non polar gases with the ionic liquids are too strong. The experimental results in room temperature ionic liquids are consistent with data on inorganic molten salts (15). In molten salts it was observed that non (quadru)polar gases like Ar, N$_2$ and He have indeed solubilities that increase with increasing temperature—endothermic solvation—whereas the solubility of CO$_2$ decreases with increasing temperature—exothermic solvation.

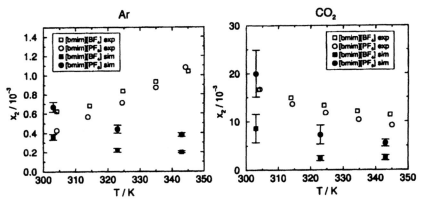

Figure 3. Temperature dependence of the solubility of Ar and CO$_2$ in the ionic liquids. The solubility corresponds to the mole fraction of dissolved gas at a partial pressure of 1 bar.

A significant aspect concerning the interactions of quadrupolar gases like CO$_2$ or even N$_2$ with the ionic liquids is the relevance of the electrostatic term. If this term is very important, then it may be necessary to include it in the description of N_2, from where it is often absent. Also, properties that depend on the unlike gas-liquid interactions will be very sensitive to details of the electrostatic part of the models. In Table I are included the solubilities calculated for CO_2 and N_2 in [bmim][BF_4] and [bmim][PF_6], in two situations: with full electrostatics, and with the partial charges set to zero. This means that a distribution of three charges was added to the model of N_2 in order to reproduce its experimental quadrupole moment (16). It is concluded that the effect of removing the quadrupole is dramatic for CO_2, causing a diminution of the solubility, whereas for N_2 an effect is detectable but not very significant. Addition of a quadrupole to the N_2 model does not change much the values of the calculated solubilities.

Table I. Effect of the quadrupole on the solubility of CO_2 and N_2 in [bmim][BF_4] and [bmim][PF_6], calculated at 303 K

$x_2/10^{-3}$	Full electrostatics	No electrostatics
	CO_2	
[bmim][BF_4]	8.6	1.3
[bmim][PF_6]	20.0	2.8
	N_2	
[bmim][BF_4]	0.12±0.02	0.10±0.01
[bmim][PF_6]	0.30±0.04	0.21±0.03

The site-site solute-solvent radial distribution functions obtained by simulation allow an analysis of the structure of the solution, and help elucidate the nature of the gas-liquid interactions. The CO_2-[bmim][PF_6] radial distribution functions (rdfs) are plotted in Figure 4. In the upper-left plot are shown the rdfs of the solute atoms with respect to the C_2 carbon in the imidazolium ring—this is the carbon atom bonded to the two nitrogens. The first peaks in the two curves, at a distance slightly above 5 Å, coincide. The situation is totally different in the lower-left plot, where the first and second peaks of the $O-C_{4,5}$ atoms surround the first peak of the $C-C_{4,5}$ rdf. The first maximum of the $O-C_{4,5}$ rdf is at a close distance of 3.5 Å. This structure means that the CO_2 molecule is closer to the C_4 and C_5 carbons, directed with its longitudinal axis towards these atoms of the ring. The CO_2 molecule is therefore not interacting strongly with the C_2, as might be expected from the Lewis-acid nature of the H_1 hydrogen which is bonded to this atom (17). In fact, an IR spectroscopy study points out that the stronger interactions of CO_2 are with the anion (11).

From the upper-right plot, containing the rdfs between the solute atoms and the P atom of the anion, the superposition of the very sharp first peaks at distances slightly above 4 Å shows that the CO_2 molecule is lying flat against the

PF_6^-. This is not unexpected owing to the positive partial charge on the C atom of CO_2.

The lower left plot contains some cation-anion atom-atom rdfs, which are included for validation purposes, since these functions can be compared to those of other studies in the literature. In particular, the present cation-anion C_2–P rdfs are in close agreement with those of Margulis, Stern and Berne (18), but show a less pronounced first peak and higher second peak than the rdfs obtained by Morrow and Maginn (19). The discrepancies between simulated results is a consequence of the diverse force fields employed. Neutron diffraction data is available for center of mass rdfs (20), so direct comparison with the present results is not straightforward.

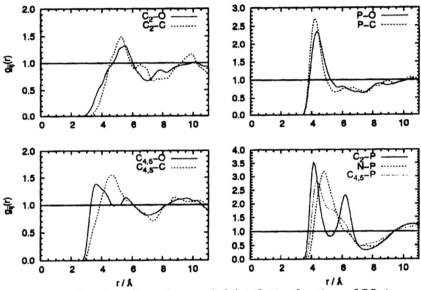

Figure 4. Site-site solute-solvent radial distribution functions of CO_2 in [bmim][PF_6] ionic liquid.

In order to assess the relevance of the electrostatic part in the interactions and structure of the solution, similar calculations as above were undertaken, but using the model for CO_2 with electrostatic charges removed. The corresponding rdfs are shown in Figure 5. It is clear, by the overlap of the first maxima of the oxygen and carbon rdfs in all plots, that the strong directional interactions of CO_2 with the imidazolium ring are now absent, and the solute molecule has much more rotational freedom. The presence of the quadrupole has a remarkable effect on the interactions of CO_2 with the ionic liquids, both in energetic and structural terms. This means that special attention should be paid to the definition of the electrostatic parts in the models of the ionic liquids and of the molecules interacting with them.

As conclusion, the present results of solubility of CO_2 in the ionic liquids show that this property can be predicted reasonably well without assuming any specific chemical interactions. Also, the structure of the solutions indicates a close interaction of CO_2 with the fluorinated anion and an absence of direct interactions between CO_2 and the H_1 proton—the site that has the strongest Lewis acid character in the cation.

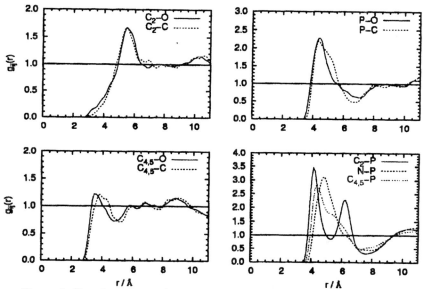

Figure 5. Site-site solute-solvent radial distribution functions of the CO_2 interaction model with the electrostatic charges removed in [bmim][PF_6].

Acknowledgment

The authors acknowledge access to the supercomputing centers in France, IDRIS of the CNRS and CINES of the Ministère de la Recherche.

References

1. Costa Gomes, M.F.; Husson, P.; Jacquemin, J.; Majer, V. *ACS Symp. Series* **2003**, this volume.
2. Canongia Lopes, J.N.; Deschamps, J.; Pádua, A.A.H. *J. Phys. Chem. B* **2003**, in press.
3. Jorgensen, W.L.; Maxwell D.S.; Tirado-Rives J. *J. Am. Chem. Soc.* **1996**, *118*, 11225 ; Kaminski, G.; Jorgensen, W. L. *J. Phys. Chem.* **1996**, *100*, 18010.
4. Cornell, W. D.; Cieplak, P.; Bayly, C. I.; Gould, I. R.; Merz, K. M.; Ferguson, D. M.; Spellmeyer, D. C.; Fox, T.; Caldwell, J. W.; Kollman, P.A. *J. Am. Chem. Soc.* **1995**, *117*, 5179. Parameters obtained from file parm99.dat corresponding to AMBER versions 1999 and 2002.
5. Michels, A.; Wijker, H.; *Physica* **1949**, *15*, 627.
6. Rivera, J.L.; Alejandre, J.; Nath, S.K.; de Pablo, J.J. *Mol. Phys.* **2000**, *98*, 43.
7. Myiano, Y. *Fluid Phase Equilibria* **1999**, *29–35*, 158.
8. Harris, J.G.; Young, K.H. *J. Phys. Chem.* **1995**, *99*, 12021.
9. Smith, W; Forrester, T.R. *The DL_POLY package of molecular simulation routines, version 2.12.* The Council for the Central Laboratory of Research Councils, Daresbury Laboratory, Warrington, UK, 1999.
10. Anthony, J.L.; Maginn, E.J.; Brennecke, J.F. *J. Phys. Chem. B* **2002**, *106*, 7315.
11. Kazarian, S.G.; Briscoe, B.J.; Welton, T. *Chem. Commun.* **2000**, 2047.
12. Blanchard, L.A.; Gu, Z.; Brennecke, J.F. *J. Phys. Chem. B* **2001**, *105*, 2437.
13. Costa Gomes, M.F.; Pádua, A.A.H. *J. Phys. Chem. B* **2003**, in press.
14. Mahoney, M.W.; Jorgensen, W.L. *J. Chem. Phys.* **2000**, *112*, 8910.
15. Field, P.E.; Green, W.J. *J. Phys. Chem.* **1971**, *75*, 821.
16. Raich J.C.; Gillis N.S. *J. Chem. Phys.* **1977**, *66*, 846.
17. Aggarwal, A.; Lancaster, L.; Sethi A.R.; Welton, T. *Green Chemistry* **2002**, *4*, 517.
18. Margulis, C.J.; Stern, H.A.; Berne, B.J. *J. Phys. Chem. B* **2002**, *106*, 12017.
19. Morrow, T.I.; Maginn, E.J. *J. Phys. Chem. B* **2002**, *106*, 12807.
20. Hardacre, C.; McMath, S.E.J.; Nieuwenhuyzen, M.; Bowron, D.T.; Soper A.K. *J. Phys. Condens. Matter* **2003**, *15*, 159.

Physical Properties

Chapter 12

Physical Property Measurements and a Comprehensive Data Retrieval System for Ionic Liquids

Joseph W. Magee[1], Gennady J. Kabo[2], and Michael Frenkel[3]

[1]Experimental Properties of Fluids Group and [3]Thermodynamics Research Center, Physical and Chemical Properties Division, National Institute of Standards and Technology, MS 838.07, 325 Broadway, Boulder, CO 80305–3328
[2]Chemistry Department, Belarusian State University, Leningradskaya 14, Minsk 220050, Belarus

NIST has initiated a collaborative project to provide key physical properties of a subset of ionic liquids that represent a selection of cations, anions and substituent chemical groups. NIST is collaborating with IUPAC task groups to establish standardized systems and to develop a comprehensive data retrieval system for ionic liquids.

By a judicious choice of its cation, anion, and substituent chemical groups, it is possible to tailor the physical properties of room-temperature ionic liquids (RTIL). Various research groups have reported physical properties for RTIL over ranges of temperature and pressure, including melting points, densities, viscosities, solubilities, liquid-liquid phase equilibria, heat capacities, etc. A broad review of this subject has recently become available (1). To date, reports of some physical properties such as thermal conductivity are quite scarce. Benchmark physical properties of high accuracy are scarcer still. For these reasons, NIST has initiated a project to provide key physical properties of a subset of RTIL that represent a selection of cations, anions and substituent chemical groups. An IUPAC task group has been formed to address these concerns and a recognized need for relevant standardized systems.

As identified at a NATO Workshop in Heraklion, Greece, a verified, web-based, public-access database of RTIL properties data is an important need of the ionic liquids community. NIST has begun to address this need by evaluating published measurements on RTIL and capturing the data into a comprehensive data retrieval system known as SOURCE. SOURCE is an archival system that presently captures more than 120 properties and contains 1.3 million data points on 17,000 different substances and 12,000 mixtures. The ionic liquids database will be built by using a record structure similar to SOURCE and maintained by the Oracle (2) relational database management system. Guided Data Capture (GDC) software will be used to speed collection of new experimental data directly from experimentalists. A second IUPAC task group has been formed to address the neeed for a comprehensive data retrieval system for RTIL.

Physical Properties

The Physical and Chemical Properties Division of NIST has developed experimental capabilities for wide-ranging, high-accuracy fluid physical properties measurements. Considering the properties types and their ranges, NIST's facilities are among the most comprehensive available in the world. It has performed experimental research on the thermodynamic and transport properties of industrially important fluids and fluid mixtures covering temperatures from 5 K to over 800 K at pressures to 70 MPa, over the full range of compositions for mixtures. It uses specially designed apparatus to measure density, heat capacity, speed of sound, phase equilibrium behavior, interfacial tension, viscosity, and thermal conductivity. Over a period of more than 40 years, the Division has used these apparatus to measure properties of cryogenic gases, natural gases, alternatives to CFC refrigerants, liquefied fuels, and aqueous systems. A program has been initiated to measure, within established limits of uncertainities, key physical properties of ionic liquids.

162

The Thermodynamics Research Center (TRC) of NIST has benefitted from a long and fruitful collaboration with the Chemistry Department of Belarussian State University (BSU), which has also developed a well equipped laboratory for thermodynamic properties measurements. Recently, a new collaboration was started between groups at NIST and at BSU to study RTIL. Samples and their chemical analyses are provided by the NIST labs, who are employing their apparatus for viscosity and thermal conductivity measurements. The BSU labs are employing their apparatus for IR spectra, density, heat capacity and Knudsen effusion measurements. The goal of our effort is to thoroughly characterize the physical properties of RTIL that represent a selection of cations, anions and substituent chemical groups. Rather than carrying out screening measurements on a lengthy list of RTIL, our goals are to limit our studies of thermodynamic and transport properties to a selection of RTIL with well-characterized chemical compositions.

The first sample selected was 1-butyl-3-methyl-imidazolium hexafluorophosphate [C$_4$mim][PF$_6$] because it is an archetypal air- and water-stable (at room temperature) RTIL. It is no coincidence that this RTIL presently has the most extensive list of published physical properties, enabling us to compare with this work and test agreement with published data within the bounds of stated uncertainties. A sample of electrochemically pure [C$_4$mim][PF$_6$] was obtained from a commercial supplier with stated impurities of less than 50 ppm of both Cl$^-$ and H$_2$O. These claims were substantiated by our own analyses of Cl$^-$ with a Cl$^-$ ion selective electrode, and of H$_2$O by Karl Fischer titration. Futhermore we did a nuclear magnetic resonance (NMR) study on the sample and also on samples that were spiked with H$_2$O and CH$_2$Cl$_2$. The NMR study established that the sample had a level of the imidazole precursor that was below the detection limit, and that the level of the solvent CH$_2$Cl$_2$ was $\approx$ 0.002 mole fraction.

Our first goal was to establish the ideal gas properties (3). No other published ideal gas data were available. We began our physical properties study with an infrared (IR) spectrum measurement. After consideration for the assignment of frequencies and an evaluation of vibrational and rotational contributions, the ideal gas thermodynamic properties of [C$_4$mim][PF$_6$] were calculated, as shown in Table I at a temperature of 298.15 K.

Table I. Dimensionless Thermodynamic Properties of [C$_4$mim][PF$_6$] in the Ideal Gas State at a Temperature T = 298.15 K, R = 8.314 472 J·mol^{-1}·K^{-1}

S° / R	C_p° / R	$(H^\circ(T)-H^\circ(0))/RT$	$-(G^\circ(T)-H^\circ(0))/RT$
56.89	22.41	14.00	42.89

Our second goal was to explore the thermodynamic behavior of the condensed phases (*4*)—crystal, glass, and ordinary liquid, and the transitions between them. A pair of calorimeters in the BSU labs were used—an adiabatic calorimeter was employed at temperatures from 5 to 311 K and a triple-heat bridge scanning calorimeter from 300 to 550 K, as shown in Table II at a temperature of 298.15 K. In this work the glass transition temperature $T_g = 190.6 \pm 0.1$ K and its associated jump in the heat capacity of 81.6 ± 1.8 J·mol^{-1}·K^{-1} were measured. In addition, the temperature of fusion $T_{fus} = 283.51 \pm 0.01$ K and the enthalpy of fusion of 19.60 ± 0.02 kJ·mol^{-1} were measured.

Table II. Dimensionless Thermodynamic Properties of [C$_4$mim][PF$_6$] in the Liquid State at a Temperature $T = 298.15$ K, $R = 8.314\ 472$ J·mol^{-1}·K^{-1}

S° / R	C_p° / R	$(H^\circ(T)-H^\circ(0))/RT$	$-(G^\circ(T)-H^\circ(0))/RT$
59.30	49.15	32.40	26.90

Planned studies in this measurement program will include other cations, anions and substituent groups and reports of the transport properties of selected RTIL.

Comprehensive Data Retrieval System

Development of a comprehensive data retrieval system encompasses a number of issues related to data submission, processing, mining, quality control, management, critical evaluation, and dissemination. It is also obvious that such a system should provide coverage for various types of data such as synthesis, catalysis, structure, manufacturing, modeling as well as thermophysical and thermochemical property data. Because of the great diversity of the data types listed above, it is envisioned that the users will have a distributed information system with access to various types of data being independently managed by different research groups and linked together via the main Web outlet.

In this section we outline major structural elements of the system currently under development which covers thermophysical and thermochemical property data of ionic liquids. We also describe how this system will be integrated into a comprehensive data retrieval system for ionic liquids.

Data Submission

Since significant interest in the measurement of the thermophysical and thermochemical properties of RTIL is a relatively new phenomenon, currently there are no internationally adopted standards for reporting these data. A significant portion of these data have been reported in journals with a limited tradition for publishing high-quality thermodynamic data and which may lack consistent requirements related to reporting of these data. These facts lead to the situation in which some of the reported data cannot be unambiguously interpreted and/or critically evaluated with regard to their uncertainties. In many instances this can drastically diminish the value of the reported numerical data for use in a variety of engineering applications. To address this problem, an IUPAC project (2002-005-1-100) (5) was initiated: "Thermodynamics of Ionic Liquids, Ionic Liquid Mixtures and the Development of Standardized Systems" (K. N. Marsh—Task Group Chairman, J. F. Brennecke, M. Frenkel, A. Heintz, J. W. Magee, C. J. Peters, L. P. N. Rebelo—members). Standardization in reporting thermodynamic data for ionic liquids can be accomplished by an expansion of the Guided Data Capture (GDC) software (6) developed by the Thermodynamics Research Center (TRC) at the National Institute of Standards and Technology (NIST).

The GDC software serves as a data-capture expert by guiding extraction of information from the data set, assuring the completeness of the information extracted, validating the information through data definition, range checks, etc., and guiding uncertainty assessment to assure consistency between compilers with diverse levels of experience. A key feature of the GDC is the capture of information in close accord with customary original-document formats. The general structure of the information to be captured is based upon answers to the following sequential queries. Each step must be complete before access to the next is possible.

(1) What is the bibliographic source?
(2) What chemical compounds were studied?
(3) What was the nature (source and purity) of the particular chemical samples?
(4) What mixtures or reactions involving the samples were studied?
(5) What properties were measured?
(6) How were the properties measured?
(7) What were the numerical values and precisions obtained?

The batch data files generated by the use of the GDC software store information in a strictly hierarchical manner. The leading record is the data source ("reference") identification with all following records associated with the identified source. The next level involves "compound," "mixture," and "reaction" specifications. For each compound, there are one or more sample descriptions including the sample source, the purification method, and the final purity, along with the method of purity determination. Samples, mixtures, and reactions have associated measured properties with numerical values. In most cases, property data sets within the batch data file consist of a "header" and a table of numerical values for state variables, properties, and uncertainties. The header includes all metadata required to define the property and experimental method used, as well as to give meaning to the table of numerical values through definition of variables and units. The compiler's main interactions with the GDC involve a navigation tree, which provides a visual representation in accord with the hierarchical structure of the batch data file as it is created.

Database Structure and Management

To create an open-access, public-domain data storage facility scoped to cover information pertaining to RTIL, an IUPAC project (7) was initiated: "Ionic Liquids Database" (K. R. Seddon—Task Group Chairman, A. Burgess, M. Frenkel, M. Gaune-Escard, A. Heintz, J. W. Magee, K. N. Marsh, R. Sheldon—members).

In this section, we discuss the structure and the management of the ionic liquids database for thermophysical and thermochemical properties (ILThermo) being developed. This database will be developed as a subsystem of the SOURCE data system (8), a comprehensive archival data system for thermophysical and thermochemical property data created and maintained by TRC at NIST. SOURCE is a large, general-purpose archive of experimental data covering thermodynamic, thermochemical and transport properties for pure compounds and mixtures of well-defined composition. The database contains numerical values for various kinds of thermodynamic and thermochemical properties of systems in all phases, and values of transport properties of fluids. It is critically important that the SOURCE system contain the estimated uncertainties for practically all of the numerical data stored. This feature allows, in principle, determination of the quality of recommended data based upon the original experimental data collected in SOURCE. The SOURCE is managed by an Oracle (2) database management system.

The Oracle server/client environment allows splitting processes between the database server and client application programs. The computer running a database server handles the database transactions while PCs running database applications concentrate upon the display and interpretation of data. At TRC, the database server, Oracle RDBMS Enterprise Edition (2), resides on a SUN 280R (2) computer running the Unix operation system, while development tools, Developer 2000 (2) suite, and other client tools reside on the NIST local network. In this configuration, client software programs run on PCs and the

166

associated server processes run on the SUN (2) machine using the NIST network and the Oracle network software, Oracle Net9 (2). Oracle software products on the server include Oracle RDBMS 9I, SQL* Plus command line, SQL Loader utility, and the Import/Export utility (2). The client software system includes a development tool suite—Developer 2000, a GUI interface server manager—Enterprise Manager, Oracle Net9 Assistant and SQL* Plus (2).

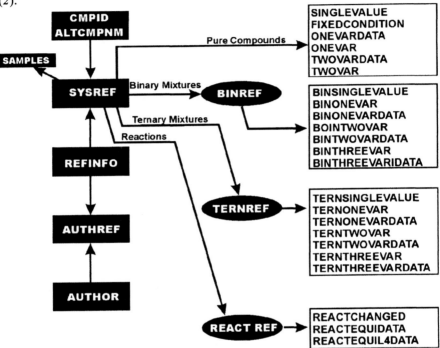

Figure 1. SOURCE database tables.

SOURCE is a relational database system. It consists of 35 tables and their relations. Figure 1 presents the relational data model for SOURCE. Each box in this figure represents one or more database tables. Tables in some boxes store the indicated information while others serve as links to relate one kind of information to another. SOURCE groups tables that contain numerical data according to the number of components—pure compounds, binary mixtures, ternary mixtures, and chemical reactions. Within each of these groups, tables express the "effective degrees of freedom" for the data sets. The effective degrees of freedom determine the number of state variables in a data record. The Gibbs Phase Rule establishes the total number of degrees of freedom for a system according to the number of independent components and number of phases. The effective number of degrees of freedom is less than the total degrees of freedom if the system contains additional constraints for a particular data set. Examples of additional constraints are: a state variable held constant, a property code that includes a constraint in the definition, or a special state such as a critical or azeotropic state. Although the Gibbs Phase Rule does not apply to transport properties, the concept of effective degrees of freedom applies to them as well.

Several ways exist to access SOURCE for input and output. These ways include a primary tool for daily data entry and maintenance—Data Entry Form on a client machine, batch input and output programs on the server machine, and SQL*Plus (2) data reports generated from both the server and the client (Figure 2).

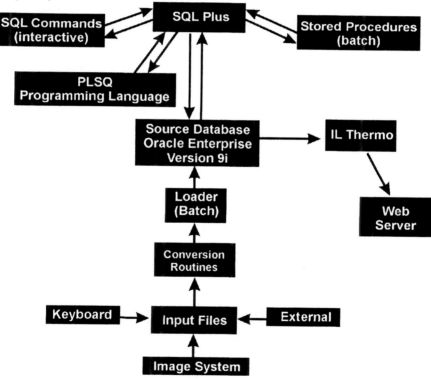

Figure 2. Database utilities (2) on the server.

Data Quality Assurance

The term "data quality," when applied to experimental-property databases, has two distinct attributes (9). The first relates to "uncertainties" assigned to numerical property values by experimentalists or professional evaluators. The second attribute refers to "data integrity," and describes the degree to which the content of the database adheres to that in the original sources and conforms to the database rules. This attribute is assessed by the extent to which stored records are complete and correct. Implementation of the mechanisms for data quality assurance includes six steps: (a) literature collection; (b) information extraction; (c) data-entry preparation; (d) data insertion into the SOURCE; (e) anomaly detection; and, (f) database rectification. The initial steps (a)-(d) can be very labor intensive and represent key components of the entire data system operation. These steps are addressed in the design of the GDC software. Steps (e) and (f) are incorporated within the data system operation.

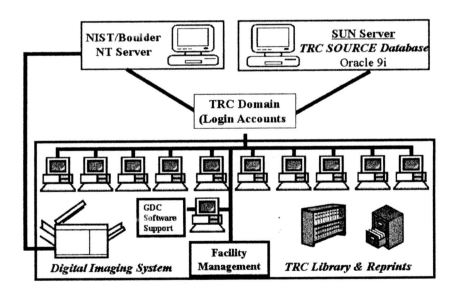

Data Collection

Personnel of the TRC Data Entry Facility will be responsible for managing all contributions to ILThermo including those from in-house compilers and from NIST collaborators worldwide. TRC operates a large in-house data-capture effort staffed chiefly by undergraduate students of chemistry and chemical engineering. The operation is supported by two networks, SUN Oracle (2) network and NIST(Boulder) network (Figure 3). Collaborators from outside NIST are involved with focused data-capture projects such as those related to specific compound types, properties, lingual sources, or contributions to the TRC Tables project. Information from original data sources is not entered directly into SOURCE, but is captured or "compiled" in the form of batch data files (coded ASCII text). This allows application of extensive checks for completeness and consistency during the capture process before the data are loaded into ILThermo. The data processing flow at the TRC Data Entry Facility with the extensive use of the GDC software is illustrated in Figure 4. The TRC Data Entry Facility is a unique facility of its kind. It operates with a data-entry rate of around 300,000 data points a year under strict guidelines for data quality assurance.

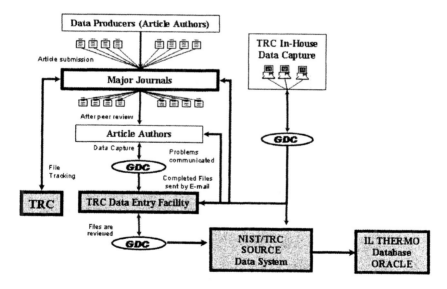

Figure 4. Data processing flow at TRC Data Entry Facility (2).

ILThermo Data Dissemination

It is currently planned to employ three different channels for ILThermo data dissemination: a Windows-based version on CD, Web-distribution of ThermoML data files, and direct Web access to the Oracle-managed data storage facility.

Windows-Based ILThermo Version

Several routes can be employed to produce versions of ILThermo that run on local servers or individual workstations. These local versions would receive periodic updates by downloading from the Oracle (2) version, as illustrated in Figure 5. The selected operational version uses the Microsoft Windows (2) operating system. Replication of a Microsoft (MS) Access (2) version from Oracle (2) databases on the server through Open Data Base Connectivity (ODBC) is a practical approach to establish a database-support system for redevelopment of a Windows (2) version of ILThermo. ODBC is a software layer that provides all databases with a common interface to communicate with each other. Microsoft Access (2) is an end-user data manipulation and query tool. MS Access (2) is familiar to most developers and users.

ThermoML Data Files

ThermoML, an XML-based approach for storage and exchange of thermophysical and thermochemical property data, has been developed in cooperation between TRC and DIPPR (Design Institute for Physical Properties) of the American Institute of Chemical Engineers (*10,11*). Batch data files created with the GDC software are converted into the ThermoML format with the software TransThermo and are then posted on the TRC Web site (*12*). ThermoML has been adopted as a prototype for an active IUPAC project (2002-055-3-024) (*13*): "XML-based IUPAC Standard for Experimental and Critically Evaluated Thermodynamic Property Data Storage and Capture" (M. Frenkel—Task Group Chairman, J. H. Dymond, E. Koenigsberger, K. N. Marsh, S. E. Stein, W. A. Wakeham—members).

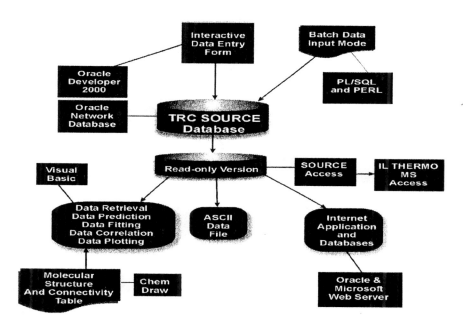

Figure 5. Production lines for TRC Window-based databases (2).

172

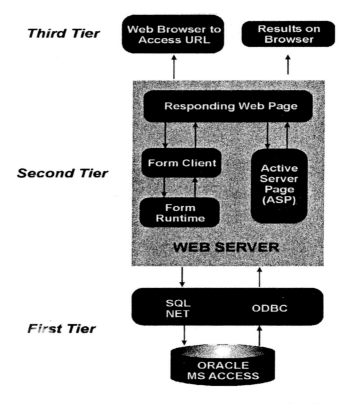

Figure 6. Three-tier web database process flow (2).

Direct Web Access to the Oracle-Managed Data Storage Facility

A three-tier rather than a two-tier architecture, shown in Figure 6, is planned to deploy Web distribution of the Oracle-based ILThermo. In this architecture, the Oracle Application Server (2) plays the role of a middle tier running on the Windows NT (2) Server platform. This architecture allows deployment of a dynamic Web site with the capacity to interact with either Oracle Databases or MS Access databases (2).

Integrated Ionic Liquids Database

It is anticipated that an integrated ionic liquids database will be developed by linking different Web servers supporting various data management operations into a central Web outlet as shown in Figure 7.

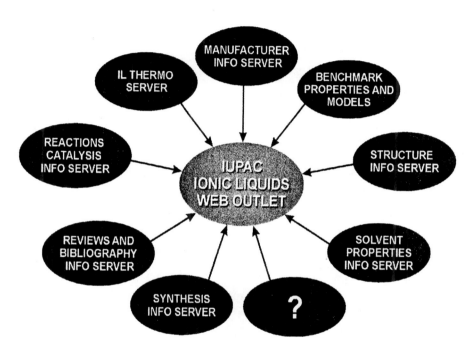

Figure 7. Distributed ionic liquids information access.

Acknowledgments

The contributions of A. V. Blohkin, R. D. Chirico, Q. Dong, Y. U. Paulechka and J. A. Widegren are gratefully acknowledged.

References

1. Wassersheid, P; Welton, T., Eds., *Ionic Liquids in Synthesis*; Wiley-VCH: Wenheim, 2003.
2. Commercial product names are provided to complete a scientific description. Such information neither constitutes nor implies endorsement by NIST or by the U.S. government. Other products may be found to serve as well.
3. Paulechka, Y. U.; Kabo, G. J.; Blohkin, A. V.; Vydrov, O. A.; Magee, J. W.; Frenkel, M. Thermodynamic Properties of 1-Butyl-3-methylimidazolium Hexafluorophosphate in the Ideal Gas State. *J. Chem Eng. Data* **2003**, *48*, 457-462.
4. Kabo, G. J.; Blohkin, A. V.; Paulechka, Y. U.; Kabo, A. G.; Shymanovich, M. P.; Magee, J. W. Thermodynamic Properties of 1-Butyl-3-methylimidazolium Hexafluorophosphate in the Condensed State. *J. Chem Eng. Data* **2004**, *49*, 453-461.
5. www.iupac.org/projects/2002/2002-005-1-100.html
6. Diky, V. V.; Chirico, R. D.; Wilhoit, R. C.; Dong, Q.; Frenkel, M. Windows-Based Guided Data Capture Software for Mass-Scale Thermophysical and Thermochemical Property Data Collection. *J. Chem. Inform. and Comp. Sci.* **2003**, *43*, 15-24.
7. www.iupac.org/projects/2003/2003-020-2-100.html
8. Frenkel, M; Dong, Q.; Wilhoit, R. C.; Hall, K. R. TRC SOURCE Database: A Unique Tool for Automatic Production of Data Compilations. *Int. J. Thermophys.* **2001**, *22*, 215-226
9. Dong, Q.; Yan, X.; Wilhoit, R. C.; Hong, X.; Chirico, R. D.; Diky, V. V.; Frenkel, M. Data Quality Assurance for Thermophysical Property Databases – Applications to the TRC SOURCE Data System. *J. Chem. Inform. and Comp. Sci.* **2002**, *42*, 473-480.
10. Frenkel, M.; Chirico, R. D.; Diky, V. V.; Dong, Q.; Frenkel, S.; Franchois, P. R.; Embry, D. L.; Teague, T. L.; Marsh, K. N.; Wilhoit, R. C. ThermoML-An XML-Based Approach for Storage and Exchange of Experimental and Critically Evaluated Thermophysical and Thermochemical Property Data. 1. Experimental Data. *J. Chem. Eng. Data* **2003**, *48*, 2-13
11. Chirico, R. D.; Frenkel, M.; Diky, V. V.; Marsh, K. N.; Wilhoit, R. C. ThermoML-An XML-Based Approach for Storage and Exchange of Experimental and Critically Evaluated Thermophysical and Thermochemical Property Data. 2. Uncertainties. *J. Chem. Eng. Data* **2003**, *48*, 1344-1359.
12. www.trc.nist.gov
13. www.iupac.org/projects/2002/2002-055-3-024.html

Chapter 13

Criticality of the [C4mim][BF4] + Water System

C. A Cerdeiriña[1], J. Troncoso[1], C. Paz Ramos[1], L. Romaní[1,*],
V. Najdanovic-Visak[2], H. J. R. Guedes[2], J. M. S. S. Esperança[2],
Z. P. Visak[2], M. Nunes da Ponte[2], and L. P. N. Rebelo[2,*]

[1]Departmento de Física Aplicada, Universidad de Vigo, Facultad de
Ciencias del Campus de Ourense, Aslagoas 32004, Ourense, Spain
[2]Instituto de Tecnologia Química e Biológica, ITQB−2, Universidade Nova
de Lisboa, Av. da Republica, Apartado 127, 2780−901, Oeiras, Portugal
*Corresponding author: romani@uvigo.es
*Corresponding author: luis.rebelo@itqb.unl.pt

A study of the behavior of the response functions of the
[C4mim][BF4] + water ionic binary solution near its liquid-
liquid critical point at atmospheric pressure is presented.
Phase equililibrium temperatures, which allow to obtain the
critical coordinates of this system, are determined.
Measurements of the isobaric heat capacity per unit volume in
the critical region indicate Ising-like behavior. The slope of
the critical line, $(dT/dp)_c$, is estimated by means of Prigogine
and Defay's equation using experimentally determined excess
volumes and excess enthalpies as a function of temperature.
$(dT/dp)_c$ is found to be near zero. The consequences of this
fact for the global critical behavior of second-order volumetric
derivatives are discussed.

Introduction

Liquid-liquid phase transitions in ionic solutions are divided into two main groups depending on the value of the dielectric constant ε of the non-ionic solvent *(1,2)*. According to the restrictive primitive model, when ε is high (solvophobic systems), Coulombic forces are weak and phase transitions are driven by specific interactions (hydrogen bonds, hydrophobic interactions, etc.). On the other hand, Coulombic forces are responsible for phase separation in low dielectric constant media (Coulombic systems). In both cases, interactions are effectively short-ranged: (*i*) long-ranged Coulombic interactions are screened by low ε and (*ii*) solvophobic interactions are always short-ranged. Within the frame of the crossover theory *(3)*, these facts mean that Ising-like behavior is expected close to the critical point, whereas, as for other complex systems, non-universal crossover behavior will be observed in the critical domain. Although no general explanation is presently accepted, it is now believed that ionic solutions exhibit crossover from Ising to some type of mean-field tricritical behavior *(2,4)*, which is not yet well understood.

Chemical instability and other experimental factors make the interpretation of the existing data a difficult task. These issues are believed to play an important role in the aforementioned incomplete understanding of ionic solutions' criticality. Further knowledge of this subject requires much more reliable data than that currently available. In this regard, the recent appearance of room temperature ionic liquids (*RTILs*) is of interest since these liquids are generally very stable and are easily controlled for purity. In addition, mixtures of *RTILs* with water and alcohols show liquid-liquid phase separation at experimentally accessible conditions. Historically, the most commonly studied family of *RTILs* is that containing 1-alkyl-3-methylimidazolium cations $[C_n\text{mim}]^+$ combined with the anions, $[BF_4^-]$ (tetrafluoroborate), or $[PF_6^-]$ (hexafluorophosphate). It is currently known that these halide-containing *RTILs* are not attractive from an industrial perspective, because they easily undergo hydrolisys in the presence of water *(5)*. Nonetheless, they still constitute model research systems, as a relatively large body of information (physical, chemical properties, and simulations) has already been accumulated. Although little is still known about the criticality of their solutions, an important body of work about phase behavior and fluctuation phenomena as well as thermodynamic-property measurements is currently being developed *(5a,6)*.

In addition to the crossover problem, there are other aspects of interest in the critical behavior of ionic solutions. Examples include the universality of critical exponents and some amplitude relationships or relations between critical amplitudes using the slope of the critical line. Such a study is presented here for the [C$_4$mim][BF$_4$] + water system at atmospheric pressure. The liquid-liquid equilibrium curve and isobaric heat capacities per unit volume C_p/V for a critical mixture in both the homogeneous and heterogeneous regions were examined. The slope of the critical line, $(dT/dp)_c$, was estimated by means of Prigogine and Defay's equation, which, under some restrictive conditions, relates this quantity to the excess volume-excess enthalpy ratio v^E/h^E at the critical point (7). To this end, experimental $v^E(T)$ and $h^E(T)$ data were determined. Combining the C_p/V data with $(dT/dp)_c$, predictions of the global critical behavior of second-order thermodynamic derivatives was made according to the framework of Griffiths and Wheeler's analysis of criticality in multicomponent systems (8).

Experimental

Chemicals

[C$_4$mim][BF$_4$] (purity > 98 %) was purchased from Solvent Innovation, and Milli-Q water was used. Prior to the measurements, vacuum was applied to the ionic liquid at 60 °C for two days owing to its hygroscopic character and possible contamination with traces of volatile compounds originated at the synthesis procedure. Then, it was placed into a dry box. Before each run, [C$_4$mim][BF$_4$] was maintained under vacuum for two hours in order to eliminate extra small traces of water.

Equipment and Techniques

Liquid-liquid phase separation temperatures were obtained by means of cloud-point determinations, where a He-Ne laser light shines through a homemade Pyrex glass cell. The whole device as well as the experimental procedure can be found elsewhere (9). Cloud-point temperatures were determined to an uncertainty of about ± 0.01 K.

Heat capacities per unit volume were determined by means of a Micro DSC II differential scanning calorimeter from Setaram. This apparatus as well as the experimental procedure (scanning method) were described elsewhere *(10)*. Isobaric runs at a scanning rate $\beta = 0.01$ K·min^{-1} were performed. The device's ability to obtain reliable data in the critical region was established in a previous work *(11)*.

Excess volumes were obtained by vibrating-tube densitometry using an Anton-Paar 512P cell *(12)*, whereas Calvet calorimetry was employed to determine the excess enthalpies *(7)*. The latter were obtained at a high single temperature, 333.15 K. This measuring temperature –usually, h^E is determined at 298.15 K– was selected owing to the large viscosity of [C$_4$mim][BF$_4$] at room temperature, a fact which produced a very long mixing process with a nearly vanishing calorimetric signal. The uncertainties in v^E and h^E are estimated to be ± 0.003 cm^3·mol^{-1} and 1 %, respectively. The procedure for obtaining h^E as a function of temperature involves the determination of excess molar isobaric heat capacities $c_p^E(T)$. They were determined combining C_p/V with density measurements. An uncertainty about ± 0.15 J·mol^{-1}·K^{-1} in c_p^E using a scanning rate of 0.25 K·min^{-1} was attained.

Results and Discussion

Coexistence Curve

Table I shows the phase separation temperatures at atmospheric pressure as a function of the ionic liquid weight fraction, w. The critical composition is located at the low-ionic liquid mole fraction region resulting in a highly unsymmetrical curve. This is partially due to the high difference between the molar volumes of the two liquids. As can be seen in Fig. 1, a nearly symmetrical curve is obtained if w is chosen as the composition variable. Data were fitted to a scaling function:

$$T = T_c\left(1 - \frac{|w - w_c|^{1/b}}{A}\right). \tag{1}$$

where T_c, w_c, A, and b are fitting parameters, the values of which are listed in Table I. The critical coordinates (w_c, T_c) for this system are 0.489 (equivalently, $x_c = 0.070$) and 277.59 K, respectively. In this particular case of the phase diagram, and due to the lack of more data points, the parameters reported in Table I should be considered as mere fitting parameters with no special significance in terms of critical phenomena.

Heat Capacity per unit Volume

Figure 2 shows isobaric heat capacities per unit volume, C_p/V, for a critical mixture as a function of temperature. The λ-curve observed is a clear indication of Ising-like behavior in the immediate proximity of the critical point, confirming the expectations already mentioned in the Introduction. Data in the homogenous region were fitted to the following expression which contains the power-law and regular terms:

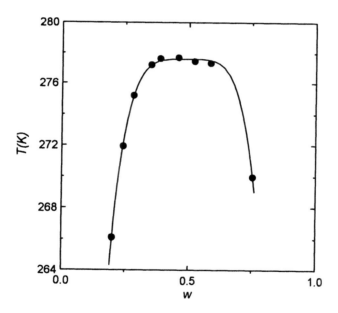

Figure 1. Liquid-liquid phase equilibrium as a T–w curve. The two-phase region is located inside the envelope.

180

Table I. Liquid–liquid phase separation temperatures, T, at atmospheric pressure for w[C$_4$mim][BF$_4$] + $(1 - w)$water mixtures and parameters of Eq.(1).

w	T/K	w	T/K
0.1992	266.11	0.4568	277.71
0.2415	271.94	0.5217	277.46
0.2819	275.20	0.5840	277.34
0.3502	277.23	0.7531	269.96
0.3854	277.65		

T_c	w_c	b	A
277.59	0.489	0.218	0.0822

$$C_p / V = B + E\left(\frac{T-T_c}{T_c}\right) + \frac{A^+}{\alpha}\left(\frac{T-T_c}{T_c}\right)^{-\alpha} \qquad (2)$$

where α and A^+ are the critical exponent and critical amplitude of C_p/V, respectively. B and E are coefficients that characterize the regular part. The fitting strategy consisted of fixing α to its universal value (0.110) allowing the remaining ones (T_c included) to vary. The results thus obtained as well as the standard deviation of the fit can be found in Table II. The value of A^+ (0.017 J·K^{-1}·cm^{-3}) lies within the typical range for molecular liquids. This is a consequence of the short-range nature of the interactions responsible for phase separation – in the present case, solvophobic interactions owing to the high ε of water. Unfortunately, the quality of the data do not permit us to obtain information about crossover behavior.

Slope of the Critical Line.

Prigogine and Defay's equation has proven to be an useful tool for predicting the value of the slope of the critical line $(dT/dp)_c$ *(7,13)*. In its simplified, restrictive version it is expressed as:

$$\left(\frac{dT}{dp}\right)_c = \lim_{\substack{x \to x_c \\ T \to T_c}} T \frac{v^E}{h^E}, \qquad (3)$$

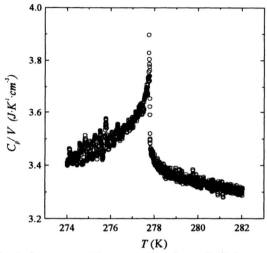

Figure 2. Isobaric heat capacities per unit volume C_p/V for a critical mixture.

Table II. Parameters of Eq. (2) and standard deviation σ.

B	E	A^+	α	T_c	σ
4.565	−8.562	0.017	0.110	277.60	0.05

Table III. Coefficients of Eq. (4) and standard deviations σ. Heat capacities in J·mol^{-1}·K^{-1} and T in Kelvin.

w	A_i			σ
	$i = 0$	$i = 1$	$i = 2$	
0.1540	−19.86	0.1198	−0.000157	0.04
0.2871	7.04	−0.0471	0.000116	0.03
0.4390	22.82	−0.1341	0.000248	0.02
0.4906	39.10	−0.2330	0.000402	0.04
0.7263	33.09	−0.1768	0.000304	0.02
0.8622	4.78	0.0021		0.04
0.9020	5.87	−0.0056		0.03
0.9420	18.23	−0.1055	0.000161	0.05
0.9772	25.96	−0.1828	0.000306	0.04

where v^E and h^E stand for excess volume and excess enthalpy, respectively.

As stated previously *(7)*, the rigorous validity of Eq. (3) requires that the curves representing v^E and h^E versus composition present a similar shape in the near-critical conditions. The h^E values at different temperatures were obtained with the aid of c_p^E measurements. The latter were fitted as a function of temperature to polynomials of the form

$$c_p^E = \sum_{i=0}^{n} A_i T^i . \tag{4}$$

The results of the fittings are given in Table III. Data at selected temperatures for v^E and h^E can be found in Table IV. The best way to explore the applicability of Prigogine and Defay's equation for the present system is by selecting the weight fraction as the composition variable as the interesting range is located at low ionic liquid mole fractions. Figure 3, where $v^E(w)$ and $h^E(w)$ curves at 298.15 K (the nearest working temperature to the critical one) are given, illustrates the fulfillment of the expected constraints.

Figure 4 shows both $v^E(T)$ and $h^E(T)$ for a near-critical composition. A quadratic temperature dependence, which is much more marked for v^E, is found in both cases. The quantity Tv^E/h^E was obtained at each temperature and, then, extrapolated to $T = T_c$, where Prigogine and Defay's equation has its highest validity. The results are reported in Table V. Taking into account that $(dT/dp)_c$ typically ranges from 2 to 20 mK·bar^{-1}, the obtained result, 0.4 mK·bar^{-1}, represents a quasi-null value. This can graphically be understood by observing the nearly vanishing v^E value at the critical point (see Fig. 4).

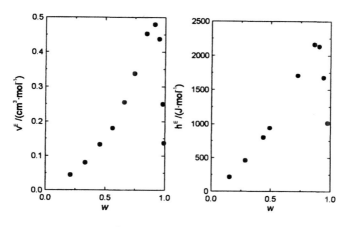

Figure 3. Excess volumes v^E and excess enthalpies h^E for w[C$_4$mim][BF$_4$] + $(1-w)$water at 298.15 K and 1 bar.

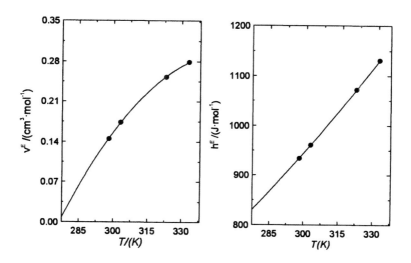

Figure 4. Excess volumes v^E and excess enthalpies h^E for a critical mixture as a function of T. (——) Second-order polynomial fit.

Table IV. Excess molar volumes, v^E, and excess molar enthalpies, h^E, for $w[C_4mim][BF_4] + (1 - w)$water as a function of temperature, T.

w	v^E (cm^3·mol^{-1})			
	$T = 298.15$ K	$T = 303.15$ K	$T = 323.15$ K	$T = 333.15$ K
0.2112	0.045	0.059	0.078	0.078
0.3344	0.081	0.100	0.147	0.156
0.4563	0.132	0.160	0.233	0.255
0.5609	0.181	0.214	0.309	0.344
0.6594	0.255	0.296	0.409	0.451
0.7459	0.338	0.382	0.510	0.561
0.8441	0.453	0.498	0.626	0.678
0.9123	0.480	0.521	0.628	0.671
0.9502	0.437	0.468	0.543	0.573
0.9817	0.251	0.255	0.259	0.273
0.9924	0.137	0.110	0.109	0.120

w	h^E (J·mol^{-1})			
	$T = 298.15$ K	$T = 303.15$ K	$T = 323.15$ K	$T = 333.15$ K
0.1540	216	225	271	296
0.2871	459	476	549	590
0.4390	797	822	924	979
0.4906	941	968	1079	1139
0.7263	1706	1743	1895	1974
0.8622	2160	2187	2295	2350
0.9020	2134	2155	2237	2277
0.9420	1677	1682	1702	1711
0.9772	1015	1008	983	973

**Table V. Experimental Tv^E/h^E for a
critical mixture as a function of T and
at atmospheric pressure.**

T (K)	Tv^E/h^E (mK·bar^{-1})
333.15	8.2
323.15	7.6
303.15	5.5
298.15	4.6
277.59	0.4*

*Extrapolated value at $T = T_c$.

Griffiths and Wheeler's geometrical picture of critical phenomena *(8)* predicts very small critical anomalies (experimentally undetectable) in the isobaric thermal expansivity α_p and the isothermal compressibility κ_T when $(dT/dp)_c$ approaches zero. Unfortunately, the κ_T critical anomaly cannot be experimentally detected even in the case of a large $(dT/dp)_c$. Furthermore, density measurements as a function of temperature would also show a very small (almost undetectable) critical anomaly for α_p. In future work, it would be desirable to determine directly the measured value of $(dT/dp)_c$ as well as check the predicted value from Prigogine and Defay's equation using experimental (not extrapolated) $v^E(w_c, T_c)$ and $h^E(w_c, T_c)$ values. Current research is being performed in our laboratories to reach this goal.

Acknowledgements

Research at the Universidad de Vigo laboratory was supported by #BFM2003-09295 and #PGIDIT-03-DPI-38301-PR. Authors are indebted to the Secretaría de Estado de Política Científica y Tecnológica (Ministerio de Ciencia y Tecnología de España) and to the Dirección Xeral de I + D (Xunta de Galicia). Work at ITQB was financially supported by Fundação para a Ciência e Tecnologia, Portugal, under contracts Nos. POCTI/EQU/34955 and POCTI/EQU/35437. V.N.-V. and J.M.S.S.E. are grateful to Fundação para a Ciência e Tecnologia for doctoral fellowships.

References

1. *Equations of State of Fluids and Fluid Mixtures (Chapter 17)*.; Sengers, J. V.; Kayser, R. F.; Peters, C. J.; White Jr, H. J. Eds.; Amsterdam; Elsevier, 2000.

2. Gutkowski, K.; Anisimov, M. A.; Sengers, J. V. *J. Chem. Phys.* **2001**, *114*, 3133.

3. Wagner, M; Stanga, O.; Schröer, W. *Phys. Chem. Chem. Phys.* **2003**, *5*, 3943.

4. (a) Albright, P. C.; Chen, Z. Y.; Sengers, J. V. *Phys. Rev. B* (Rapid Comm.) **1987**, *36*, 877. (b) Anisimov, M. A.; Povodyrev, A. A.; Kulikov, V. D.; Sengers, J. V. *Phys. Rev. Lett.* **1995**, *75*, 3146. (c) Melnichenko, Y. B.; Anisimov, M. A.; Povodyrev, A. A.; Wignall, G. D.; Sengers, J. V.; Van Hook, W. A. *Phys. Rev. Lett.* **1997**, *79*, 5266.

5. (a) Najdanovic-Visak, V.; Esperanca, J. M. S. S.; Rebelo, L. P. N.; Nunes da Ponte, M.; Guedes, H. J. R.; Seddon, K. R.; Szydlowski, J. *Phys. Chem. Chem. Phys.* **2002**, *4*, 1701; (b) Visser, A. E.; Swatloski, R. P.; Reichert, W. M.; Griffin, S. T.; Rogers, R. D. *Ind. Eng. Chem. Res.* **2000**, *39*, 3596; (c) Swatloski, R. P.; Holbrey, J.D.; Rogers, R. D. *Green Chem.* **2003**, *5*, 361.

6. (a) Najdanovic-Visak, V.; Esperanca, J. M. S. S.; Rebelo, L. P. N.; Nunes da Ponte, M.; Guedes, H. J. R.; Seddon, K. R.; de Sousa, H. C.; Szydlowski, J. *J. Phys. Chem. B* **2003**, *107*, 12797, and references therein. (b) Najdanovic-Visak, V.; Serbanovic, A.; Esperanca, J. M. S. S.; Guedes, H. J. R.; Rebelo, L. P. N.; Nunes da Ponte, M. *Chem. Phys. Chem.* **2003**, *4*, 520. (c) Rebelo, L. P. N.; Najdanovic-Visak, V.; Gomes de Azevedo, R.; Nunes da Ponte, M.; Guedes, H. J. R.; Visak, Z. P.; de Sousa, H. C.; Szydlowski, J.; Canongia Lopes, J. N.; Cordeiro, T. C.; this same book, and references therein.

7. Rebelo, L. P. N.; Najdanovic-Visak, V.; Visak, Z.P.; Nunes da Ponte, M.; Troncoso, J.; Cerdeiriña, C.A.; Romaní, L. *Phys. Chem. Chem. Phys.* **2002**, *4*, 2251.

8. Griffiths, R. B.; Wheeler, J. C.; *Phys. Rev. A* **1970**, *2*, 1047.

9. de Sousa, H. C.; Rebelo, L. P. N. *J. Chem. Thermodyn.* **2000**, *32*, 355.

10. Cerdeiriña, C. A.; Miguez, J. A.; Carballo, E.; Tovar, C. A.; de la Puente, E.; Romaní, L. *Thermochim. Acta* **2000**, *347*, 37.

11. Cerdeiriña, C. A.; Troncoso, J.; Carballo, E.; Romaní, L. *Phys. Rev. E* **2002**, *66*, 031507.

12. Gomes de Azevedo, R.; Szydlowski, J.; Pires, P.F.; Esperança. J.M.S.S.; Guedes, H.J.R.; Rebelo, L.P.N. *J. Chem. Thermodyn.* **2004**, *36*, 211.

13. Myers, D. B.; Smith, R. A.; Katz, J.; Scott, R. L.; *J. Phys. Chem.* **1966**, *70*, 3341.

Chapter 14

Activity Coefficients and Heats of Dilution in Mixtures Containing Ionic Liquids

Andreas Heintz[1], Wojciech Marczak[2], and Sergey P. Verevkin[1]

[1]Department of Physical Chemistry, University of Rostock, D–18055 Rostock, Germany
[2]Institute of Chemistry, University of Silesia, PL–40006, Katowice, Poland

Vapor-liquid equilibria of the mixtures of high boiling solvents (aldehydes, ketones, esters, ethers, amines) with the ionic liquid 1-methyl-3-ethyl-imidazolium bis(trifluoromethyl-sulfonyl) imide [emim][ntf$_2$] were studied by using the transpiration method, which allows to determine the composition of the vapor phase of high boiling mixtures and to calculate activity coefficients in the liquid phase. The measurements were carried out over the whole concentration range at different temperatures between 298 K and 323 K. Enthalpies of solution of 6 organic solutes in the [emim][ntf$_2$] have been measured at 298.15 K in the range of low concentrations using titration calorimetry. Results at infinite dilution are compared with indirectly obtained data from activity coefficients in infinite dilution.

The unique properties of ionic liquids open up a wide range for various applications. Ionic liquids have proved to be viable reaction media for numerous types of reactions. Separation of the reaction products from the ionic liquid requires the knowledge of the thermodynamic properties of such mixtures as well as of properties of the pure individual compounds. This work continues our study of thermodynamic properties of mixtures containing ionic liquids [1-7] Activity coefficients and data of mixing enthalpies, which are needed for many design calculations, are still unavailable for systems containing ILs. Vapor-liquid equilibrium of the mixtures of high boiling solvents (aldehydes, ketones, esters, ethers, amines) with the ionic liquid [emim][ntf$_2$] - 1-methyl-3-ethyl-imidazolium bis(trifluoromethyl-sulfonyl) imide ($C_8H_{11}S_2O_4F_6N_3$) were studied by using the transpiration method. The measurements were carried out over the whole concentration range at different temperatures between 298 K and 323 K. Activity coefficients γ_i of these solvents in the ionic liquid have been derived and from their temperature dependence partial molar excess enthalpies H_i^E of the solutes in the ionic liquid have been estimated.

1. Activity coefficients of high boiling solutes in ionic liquid in the whole concentration range

Vapor-liquid equilibrium experiments carried out by dynamic recirculation stills, static method and headspace methods have traditionally been used to obtain information about activity coefficients in the liquid phase and a substantial body of data [8] has been accumulated in this area on the solution behaviour of a large number of mixtures. However all these traditional methods have generally been applied for investigating relatively low boiling compounds or providing results for high-boiling compounds at elevated temperatures above 373 K. For the development of separation technologies the knowledge of thermodynamic properties of the mixtures containing high-boiling compounds with with ILs at ambient temperatures is of the crucial importance. One of the suitable methods used for this purpose is the so-called transpiration method. This method is based on using an inert carrier gas stream being loaded with the equilibrium vapor of the vapor-liquid system. When the volume and the molar mass of the entrained vapor and appropriate amount of the carrier gas is known, the vapor pressure in the system can be derived from the ideal gas law provided the vapor pressure is low enough. The method has been successfully applied in our laboratory [9] for measuring vapor pressures of pure compounds and has proved to give results which are in excellent agreement with other established techniques for determining vapor pressures in the range of 0.005 to

10000 Pa. Taking into account that ILs have vanishing vapor pressures, we decided to apply this method for the investigation of the mixtures containing a high-boiling compound (solute) and an IL (solvent). In this case, knowing the starting composition of the liquid phase x_i, the composition of the vapor phase y_i is governed only by the volatile solute. By isothermal measurements of compositions of the vapor phase and by screening the compositions of the liquid phase, *i.e.* mole fraction of the solute in the mixture, and additionally having established the vapor pressure of the pure solute activity coefficients in the mixture of solute in IL can be derived. From the temperature dependence of activity coefficients partial molar heats of solution are obtained. Details of the technique can be found elsewhere (4,7). Values of γ_{solute} have been obtained according to

$$\gamma_{solute} = \frac{P}{P_{io}} \cdot \frac{1}{x_{solute}} \tag{1}$$

where P is the measured pressure of the solute at the mole fraction x_{solute} and P_{io} is the vapor pressure of the pure solute, this is the value of P at $x_{solute} = 1$. Test measurements including thermodynamic consistency tests with the mixture (n-pentanol + decane) have been made showing excellent agreement of the results of γ_i with literature data obtained by a different method (10). Measurements of γ_{solute} covering the whole range of concentration of solute + ionic liquid mixtures have been performed. A series of aldehydes, ketones, esters, ethers, amines mixed with ionic liquid has been studied. As an example Fig. 1 and Fig 2. show the results of nonanal where the pressure as well as the activity coefficients γ_{solute} of the solute are presented as function of the mole fraction of the solute.

2. Enthalpies of solution of organic solutes in the ionic liquid [emim][ntf$_2$]

Recently we have reported measurements of activity coefficients at infinite dilution γ_i^∞ of 38 solutes at different temperatures (1-3). The temperature dependence of γ_i^∞ allows to determine enthalpies of solution at infinite dilution $H_i^{E\infty}$ according to

$$\frac{H_i^{E\infty}}{R} = \left(\frac{\partial \ln \gamma_i^\infty}{\partial T^{-1}} \right)_p . \tag{2}$$

In this work, values of the heat of dilution, defined as partial molar excess enthalpy H_i^E of a series of solutes i (methanol, t-butanol, 1-hexanol, chloroform, toluene, and ethylene glycol) in the ionic liquid: [emim][ntf$_2$] were

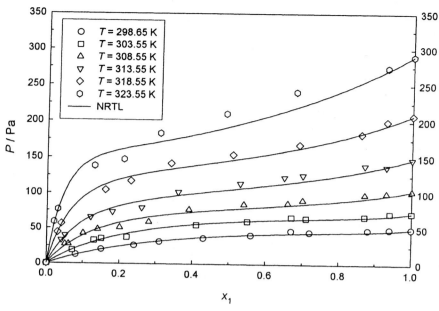

Figure 1. Partial pressure data of nonanal in the mixture with [emim][ntf₂] as function of x₁ (nonanal)

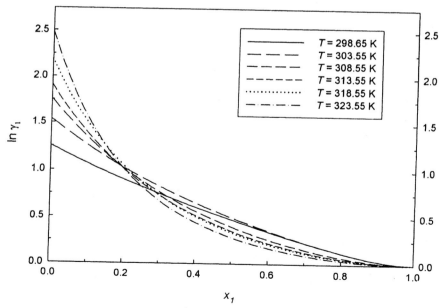

Figure 2. Values of lnγ₁ of nonanal in the mixture with [emim][ntf₂] as function of x₁ (nonanal)

measured at 298.15 K by using a titration calorimeter 2277 Thermal Activity Monitor (Thermometrics, Sweden). Extrapolation of values of H_i^E as function of solute concentration to infinite dilution allows to determine $H_i^{E\infty}$ and to make a comparison with indirectly determined values of $H_i^{E\infty}$ obtained by Eq. (2). The heat effect of the solute injection into the ionic liquid (Q_i) was recalculated into the molar enthalpy of solution (*i.e.* the partial molar excess enthalpy of the solute, H_i^E) by the following formula:

$$H_i^E = \left(\frac{\partial H^m}{\partial n_i} \right)_{T,p,n_{IL}} \approx \frac{Q_i}{\Delta n_i},$$

(3)

where n_i and n_{IL} (given in moles) denote the content of the solute and the ionic liquid in the solution, respectively. H^m is the enthalpy of mixing and Δn_i is the drop size (in moles). The experimental procedure was described elswere (11). Results are presented in Table 1.

Comparison of heats of solution of different solutes in the ionic liquid [emim][ntf₂] obtained by a direct calorimetric method could be made with values for the same systems obtained by the indirect method and already presented previously where the temperature dependence of activity coefficients of the solute in the ionic liquid was used. Considering experimental errors involved into both methods the agreement is satisfactory even though thermodynamic consistency was not confirmed in all cases. This indicates that the estimation of experimental errorsmade in ref. (1-3) was too low.

Conclusions

Systematic investigation of activity coefficients in mixtures containing ionic liquids allow to develop reliable methods of predicting solubilities of gases and vapors in ionic liquid where direct experimental data are not available. The knowledge of heats of mixing enable scientists working on the application of ionic liquids in chemical and separation processes to determine the temperature dependence of solubility data.

Table 1. Limiting partial molar excess enthalpies $H_i^{E\infty}$ of organic solutes in [emim][ntf$_2$] at 298.15 K obtained calorimetrically (this work) and by gas-liquid chromatography (1-3)

Solute	$H_i^{E\infty}$ (J·mol^{-1})			
	This work		Literature (1-3)	
methanol	5639	± 54	7787	± 360
t-butanol	7039	± 47	7240	± 340
1-hexanol	10076	± 616	10812	± 450
chloroform	−3255	± 15	−1722	± 50
toluene	−1018	± 8	−683.5	± 31
ethylene glycol	11535	± 200		

References

(1)   Heintz, A.; Kulikov, D. V.; Verevkin, S. P. *J. Chem. Eng. Data* **2001,** *46,* 1526-1529.

(2)   Heintz, A.; Kulikov, D. V.; Verevkin, S. P. *J. Chem. Eng. Data* **2002,** 47, 894-899.

(3)   Heintz, A.; Kulikov, D. V.; Verevkin, S. P., *J. Chem. Thermodyn.* **2002,** *34,* 1341-1347

(4)   Verevkin, S. P.; Vasiltsova, T. V.; Bich, E.; Heintz, A., *Fluid Phase Equil.* **2004,** *218,* 165-175.

(5)   Heintz, A.; Lehmann, J. K.; Wertz, C. *J. Chem. Eng. Data* **2003,** *48,* 472 – 474.

(6)   Heintz, A.; Klasen, D.; Lehmann, J.K. *J. Sol. Chem.* **2002,** *31,* 467-476.

(7)   Heintz, A.; Lehmann, J. K.; Verevkin, S. P. Thermodynamic properties of Liquid Mixtures Containing Ionic Liquids in "*Ionic Liquids as Green Solvents: Progress & Prospects,* "ACS Symposium Series 856, American Chemical Society, Washington DC, **2003,** 134-151.

(8)   Gmehling, J.; Onken, U.; Arlt, W. *Vapor-Liquid Equilibrium Data Collection.* DECHEMA, Frankfurt, **1977.**

(9)   Kulikov, D. V.; Verevkin, S. P.; Heintz, A. *Fluid Phase Equil.* **2001,** *192,* 187-202.

(10) Treszczanowicz T.; Treszczanowicz, J. *Bull. Acad. Pol. Sci.* **1979,** *27,* 689.

(11) Marczak, W.; Verevkin, S. P.; Heintz, A. *J. Sol. Chem.* **2003,** *32,* 519-526.

Chapter 15

Thermal and Kinetic Studies of Trialkylimidazolium Salts

Douglas M. Fox[1,*], Walid H. Awad[2], Jeffrey W. Gilman[2], Paul H. Maupin[3], Paul C. Trulove[4], and Hugh C. De Long[4]

[1]Chemistry Division, Naval Research Laboratory, Washington, DC 20375
[2]Fire Research Division, Building and Fire Research Laboratory, National Institute of Standards and Technology, Gaithersburg, MD 20899
[3]Office of Basic Energy Sciences, Office of Science, U.S. Department of Energy, Washington, DC 20585
[4]Directorate of Chemistry and Life Sciences, U.S. Air Force Office of Scientific Research, Arlington, VA 22203
*Corresponding author: telephone: 202–404–6356; fax: 202–767–3321; email: fox@nrl.navy.mil

The thermal properties of trialkylimidazolium room temperature ionic liquids (RTILs) have been determined using a flashpoint apparatus and the technique of thermal gravimetric analysis (TGA). Using a Setaflash flashpoint analyzer, all the salts studied were found to have flashpoints above 200°C. TGA was utilized to study the decomposition of imidazolium based RTILs. The effects of C-2 hydrogen substitution, structural isomerism, alkyl chain length, anion type, and purge gas type is discussed. The decomposition kinetics of 1,2-dimethyl-3-butylimidazolium hexafluorophosphate was investigated using TGA. The global kinetic model and TGA Arrhenius parameters have been determined by employing both constant heating rate and isothermal programs.

Introduction

Room temperature ionic liquids (RTILs) have received considerable attention over the past decade. Two key properties that make these compounds attractive are their relatively high thermal stability and apparent nonflammability. These characteristics are particularly important for the application of RTILs in clay-in-polymer nanocomposites, thermal fluids, and industrial processes.[1-3] Although early studies have suggested very high thermal stabilities,[4,5] recent work by Van Valkenburg et. al.[3] and Fox et. al.[6] have indicated that the long term thermal stability may not be high enough for some applications. In addition, there have been numerous reports in literature claiming that imidazolium based RTILs are nonflammable despite the lack of quantifiable data.

In this study, the flammability and thermal stability of imidazolium based RTILs are investigated. The effects of C-2 hydrogen substitution, structural isomerism, alkyl chain length, and anion type on the thermal stability are summarized and discussed. The flammabilities of the RTILs are predicted using a correlation to thermogravimetric data and verified for two bromide salts using a flashpoint apparatus. Finally, the TGA decomposition kinetics of 1-butyl-2,3-dimethylimidazolium hexafluorophosphate is analyzed using an isothermal model-based approach and an isoconversional model-free approach. The nomenclature used in this work is listed in Table 1.

Experimental

RTIL Preparation

For this study, the imidazolium based RTILs were prepared in acetonitrile, as previously described.[6] Elemental analysis was performed by Galbraith Laboratories, resulting in mass errors in carbon analysis $\leq$ 1.2%, hydrogen analysis $\leq$ 2.9%, and nitrogen analysis $\leq$ 2.1% for all the salts used in this study.

Table 1: Nomenclature used in this study

Abbreviation	Definition
BMI	1-Butyl-3-methylimidazolium
HdMI	1-Hexadecyl-3-methylimidazolium
DMPI	1,2-Dimethyl-3-propylimidazolium
DMBI	1-Butyl-2,3-dimethylimidazolium
DMiBI	1-iso-Butyl-2,3-dimethylimidazolium
DMDI	1-Decyl-2,3-dimethylimidazolium
DMHdI	1-Hexadecyl-2,3-dimethylimidazolium
TFSI	Bis(trifluoromethanesulfonyl)imide

Instrumentation

Flashpoints were measured using a Setaflash, closed cup apparatus with a maximum operating temperature of 200°C.[‡] Thermal stabilities were determined using a TA Instruments, Hi-Res TGA2960 Thermogravimetric Analyzer. For the comparative TGA study, 4.0 mg to 6.0 mg samples were placed in open ceramic pans and heated at a scan rate of 10°C/min while purged with 100 ml/min N_2. For the isothermal TGA study, 5.0 ± 0.1 mg samples of DMBIPF$_6$ were heated as quickly as possible without exceeding the final temperature. For the isoconversional TGA study, 5.0 ± 0.1 mg samples of DMBIPF$_6$ were heated at scan rates of 6°C/min, 8°C/min, 10°C/min, and °C/min. For all TGA anlyses, the mean of three replicate measurements was typically reported. The temperature of both the onset (5% mass fraction loss) and peak mass loss rate have an uncertainty of $\sigma = \pm 4$°C. The mass loss rate has a relative uncertainty of $\sigma/\bar{x} = \pm 10\%$.

Results and Discussion

Flashpoints and Thermal Stability

The measured flashpoints of a representative set of imidazolium based RTILs were compared to the literature flashpoints of several conventional polar

---

[‡] The authors wish to thank the scientists at Occupational Safety and Health Administration – Salt Lake Technical Center for performing the flashpoint analyses.

aprotic solvents and some common "green solvents", such as ethyl lactate and terpenes.[6] No flashpoints were detected for any of the RTILs used in this study below the 200°C limit of the instrument. As indicated in Figure 1, the imidazolium based RTILs exhibit flashpoints at least 100°C higher than conventional, polar, aprotic solvents and over 150°C higher than other "green solvents."

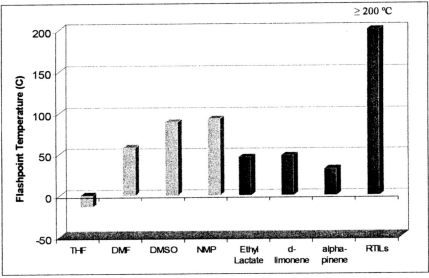

Figure 1. MSDS reported flashpoints for conventional polar, aprotic solvents and some common green solvents and measured flashpoints for imidazolium based RTILs

Lyon has recently shown that the flashpoints of a variety of organic species are approximately equal to the decomposition temperatures in N_2, where the decomposition temperature is the mean of the onset and peak decomposition temperatures determined by thermogravimetric analysis (TGA).[7] This relation is shown in Figure 2. Assuming this equality holds for RTILs, then the flashpoints of imidazolium based RTILs can be estimated from TGA decomposition data. The decomposition onset temperature, peak decomposition temperature, and estimated flashpoint temperatures for the RTILs used in this study are shown in Table 2. The estimated flashpoint temperatures indicate that the imidazolium based RTILs would offer at least a 200°C improvement in flashpoint over current green solvents and trialkylimidazolium RTILs with the more stable anions (such as BF_4^- and TFSI⁻) would most likely offer a 350°C improvement in the flash point. We are currently verifying this assertion using a Cleveland Open Cup apparatus with a higher operating temperature. Preliminary results for the bromide salts

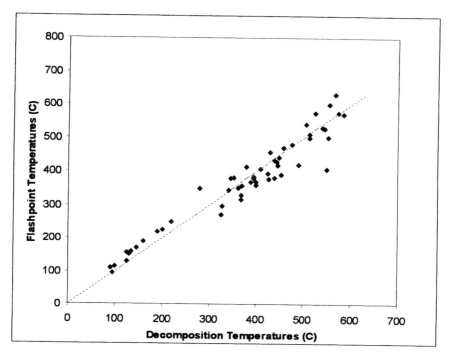

Figure 2. Correlation between flashpoint temperatures and decomposition temperatures in N_2 [6] – (Reproduced with permission from Reference 6. Copyright 2003 The Royal Society of Chemistry.)

(BMIBr = 224°C and DMBIBr = 234°C) do indeed suggest this correlation extends to imidazolium based RTILs.

The effects of C-2 hydrogen substitution, structural isomerism, alkyl chain length, and anion type on the thermal stability were examined using thermogravimetric analysis. As indicated in Figure 3, replacing the reactive C-2 hydrogen with a methyl group results in a 20-40°C improvement in the anaerobic thermal stability.

The effects of structural isomerism and alkyl chain length on the thermal stability are shown in Figure 4. When the RTILs are decomposed in air, structural isomerism has a negligible effect on the thermal stability, and increasing the alkyl chain length decreases the thermal stability. This is consistent with an aliphatic decomposition mechanism.[2] In the absence of oxygen, changing the alkyl group from n-butyl to i-butyl lowers the thermal stability by about 50°C, and increasing the alkyl chain length does not greatly affect the thermal stability. This indicates decomposition occurs through a

Table 2: TGA Decomposition Temperatures for RTILs

RTIL	$T_{onset}^{2,6}$ (°C)	$T_{peak}^{2,6}$ (°C)	$T_{decomp} \sim T_{fp}$ (°C)
BMICl	234	285	260
BMIPF$_6$	341	420	398
HdMICl	230	292	261
DMPIPF$_6$	385	493	439
DMBICl	255	300	283
DMBIBr	265	326	304
DMBISCN	265	316	295
DMBIBF$_4$	416	487	460
DMBITFSI	421	491	467
DMBIPF$_6$	403	495	449
DMiBIPF$_6$	382	439	411
DMDIPF$_6$	383	484	440
DMHdICl	239	292	265
DMHdIPF$_6$	390	470	430

Note: T_{onset} and T_{peak} data taken directly from references 2 and 6.

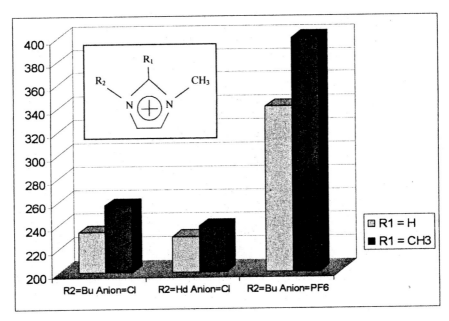

Figure 3. Effect of C-2 hydrogen substitution on TGA decomposition onset temperatures in N_2

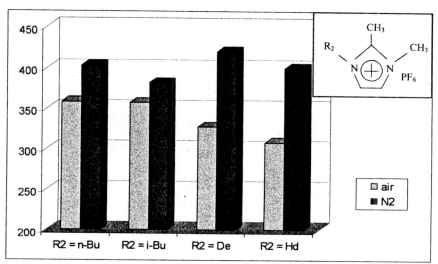

Figure 4. Effect of alkyl chain length on TGA decomposition onset temperatures

different mechanism than when oxygen is present. Indeed, the pyrolysis of RTILs has been proposed to decompose via a homolytic mechanism.[2]

Figure 5 illustrates the effects of anion type on the decomposition of DMBI compounds. The presence of a nucleophilic anion, such as SCN⁻ or a halide ion, reduces the thermal stability by over 100°C and evidence suggests decomposition proceeds by S_N1 or S_N2 nucleophilic decomposition.[2]

Kinetic Studies

It has been established that the imidazolium based RTILs exhibit excellent short term stability. However, Van Valkenburg et. al. have suggested that prolonged exposure to temperatures below the determined onset temperatures will result in significant decomposition of the RTILs.[3] To further investigate this effect, we measured the isothermal decomposition rate for DMBIPF₆ at several temperatures (*cf.* Table 3). Although the onset temperature determined by nonisothermal TGA was over 400°C, the isothermal experiments clearly show that decomposition occurs at much lower temperatures. Indeed, at 300°C, the mass loss becomes quite significant, with a mass loss rate equivalent to 15%/hr. There appears to be some decomposition at temperatures as low as 200°C. However, the mass loss rate at this temperature is close to the error limit of the experiment and any decomposition that does occur at this temperature is likely due to the loss or decomposition of the contaminants, such as water, DMBIBr, and NH₄PF₆.

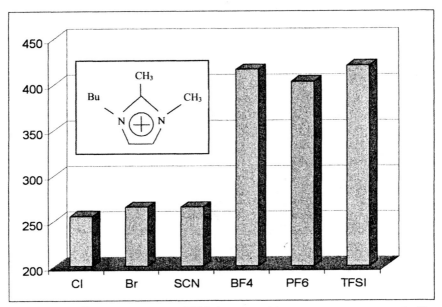

Figure 5. Effect anion type on TGA decomposition onset temperatures in N_2

By analyzing the decomposition kinetics, one can obtain a more complete and concise picture of the thermal stability. Although kinetic analysis by thermogravimetric methods does not provide information on specific chemical reactions and mechanisms, it is useful for determining global kinetic models that describe complex "pseudo species." Furthermore, as Bourbigot and co-workers[8] correctly point out, TGA measures the rate of evaporation of degradation products and not the intrinsic chemical reactions; thus, both physical and chemical processes influence the measured rate of mass change. Nevertheless, by determining an appropriate global kinetic model and calculating the corresponding Arrhenius parameters, the mass loss rate at any temperature or any heating rate program can be predicted.

Table 3: Isothermal Decomposition of DMBIPF₆

Temperature	100	200	250	300	310	320	330	340	350	360
Decomposition Rate (%/min)	0	0.01	0.03	0.25	0.37	0.63	0.96	1.28	2.18	2.59

Most decompositions are heterogeneous reactions, involving several steps, such as nucleation, growth, ingestion, and evaporation. Various sigmoid, geometric, reaction order, and diffusion-based models have been derived, and all have the form described in equation (1),

$$f(\alpha) = \alpha^m (1-\alpha)^n [-\ln(1-\alpha)]^p = \frac{1}{k(T)} \frac{d\alpha}{dt}$$ (1)

where m, n, and p are constants fit to experimental TGA data. The integral form of equation (1) is often more useful, and is shown in equation (2).

$$g(\alpha) = \int_0^\alpha \frac{d\alpha}{f(\alpha)} = k(T) * t$$ (2)

One approach to determining the kinetic parameters is to plot the integral form of the kinetic models, g(α), versus the corrected time (time minus the induction period), for several isothermal decomposition data sets. Assuming the Arrhenius parameters are constant over the entire conversion range, α, the model producing the most linear plot is considered the most appropriate model.[9] The value of the reaction rate constant, k(T), and its standard deviation, s_b, for each model may be determined from the slope of the plots. An example of this for the decomposition of DMBIPF$_6$ using the 4th order Avrami-Erofeev model is shown in Figure 6.

Then, the Arrhenius parameters can be determined from a conventional Arrhenius plot, as illustrated in Figure 7. The results for the isothermal decomposition of DMBIPF$_6$ using several global kinetic models are shown in Table 4. The best fit model is the 4th order Avrami-Erofeev, and this model is consistent with the homolytic mechanism proposed earlier.

There are several disadvantages to using isothermal, model-based analysis of decomposition data. First, as indicated in Table 4, the Arrhenius parameters vary significantly depending on the reaction model chosen. Second, the decomposition usually proceeds by multiple mechanisms, so using one model over the entire conversion range is often inappropriate. Third, many organic compounds yield significant amounts of char, which can change the decomposition rate or lower the sample temperatures as the reaction proceeds. Finally, even the best fitting models reveal little about the decomposition mechanism.

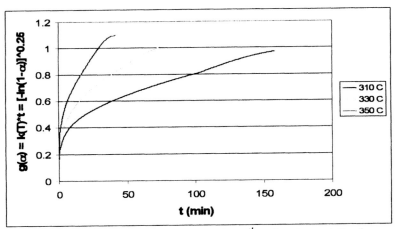

Figure 6. Determination of rate constant using 4th order Avrami-Erofeev model

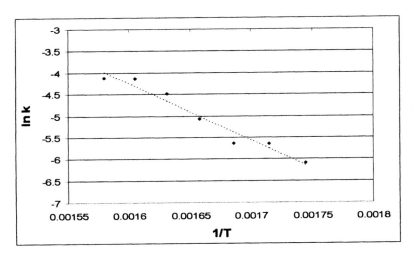

Figure 7. Arrhenius plot for $DMBIPF_6$ TGA decomposition reaction

Table 4: Arrhenius Parameters for Isothermal Decomposition of DMBIPF$_6$

Model	g(α)	Ea (kJ/mol)	A (min^{-1})
Power Law	$\alpha^{1/4}$	98	1.61 x 10^6
Power Law	$\alpha^{1/2}$	107	1.26 x 10^7
Avrami-Erofeev	$[-\ln(1-\alpha)]^{1/4}$	109	1.71 x 10^7
Avrami-Erofeev	$[-\ln(1-\alpha)]^{1/3}$	115	7.65 x 10^7
Mampel (1st order)	$-\ln(1-\alpha)$	188	2.90 x 10^{14}
Contracting sphere	$1-(1-\alpha)^{1/3}$	150	3.48 x 10^{10}
Contracting cylinder	$1-(1-\alpha)^{1/2}$	149	3.78 x 10^{10}

The TGA Arrhenius parameters can also be derived model-free, using one of the isoconversional methods. Three of the most commonly used isoconversional methods are the integral method of Ozawa, Flynn, and Wall,[10,11] the derivative method of Friedman,[12] and the non-linear method of Vyazovkin.[13] For this study, we utilized the Ozawa-Flynn-Wall method. The natural logarithm of the heating rate versus the inverse of the temperature at a specific conversion rate is plotted for DMBIPF$_6$ in Figure 8. The slopes of the lines are equal to $-E_a*b/R$, where b is a constant and its value is dependent on the value of E_a at the given conversion. The final value of the activation energy at each conversion rate is found iteratively using the method developed by Flynn.[14] The pre-exponential factor, A, can be found from equation (3), where the temperature integral is typically evaluated using an approximation, such as the Coats-Redfern equation.[15]

$$g(\alpha) = \frac{A}{\beta} \int_0^{T_\alpha} \frac{-E_a}{RT} dT \qquad (3)$$

We have calculated the activation energies at several conversion rates (cf. Table 5), but due to some recently reported misgivings of the Coats-Redfern equation, we have not used equation 3 to determine the pre-exponential factors for the various models.[16]

Table 5: Isoconversional TGA Arrhenius Parameters for DMBIPF$_6$

Conversion (%):	2	5	10	15	20	25	50	75
E_a (kJ/mol):	50	75	97	104	109	112	120	111

As indicated in Table 5, the activation energy increases from 50 kJ/mol at 2% conversion to ~110 kJ/mol at 20% conversion, then remains fairly constant during the remainder of the decomposition. This final value is in agreement with the result obtained from isothermal data using the Avrami-Erofeev models. The lower values of the activation energy at lower conversion rates is likely due to thermal lag, the production and build-up of char, or the presence of contaminants. We are currently conducting a systematic study of the impurity effects on the thermal decomposition of $DMBIPF_6$.

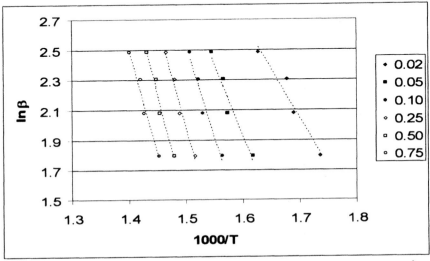

Figure 8. Determination of Ea for DMBIPF₆ TGA decomposition using the Ozawa-Flynn-Wall integral method

Conclusions

The thermal study on imidazolium based RTILs presented here illustrates the thermal effects of C-2 hydrogen substitution, structural isomerism, alkyl chain length, anion type, and purge gas type. No flashpoints below 200°C were detected for any RTILs used in this study. Flashpoints for all RTILs were predicted using a relation to TGA decomposition temperatures and verified for two bromide salts. The flashpoints for imidazolium based RTILs with nucleophilic anions are likely to be around 250°C, while imidazolium based RTILs with larger fluoride containing anions are likely to be closer to 400°C. The decomposition onset temperatures of trialkylimidazolium salts in nitrogen were greater than the analogous dialkylimidazolium salts, and were found to be more dependent upon the anion than the cation. In addition, it was found that although the calculated onset temperatures were above 350°C, significant decomposition does occur 100°C or more below these temperatures. The best fit global kinetic model using isothermal TGA data was found to be the Avrami-Erofeev (4th order) with Ea ~ 110 kJ/mol and A ~ 1.7×10^7 min^{-1}. The activation energy using constant heating rate TGA and an isoconversional approach was determined to be ~ 110 kJ/mol.

References

* - The policy of the National Institute of Standards and Technology (NIST) is to use metric units of measurement in all its publications, and to provide statements of uncertainty for all original measurements. In this document however, data from organizations outside NIST are shown, which may include measurements in non-metric units or measurements without uncertainty statements. The identification of any commercial product or trade name does not imply endorsement or recommendation by NIST or the United States Air Force (USAF). Opinions, interpretations, conclusions, and recommendations are those of the authors and are not necessarily endorsed by the USAF or NIST.

1. J. W. Gilman, W. H. Awad, R. D. Davis, J. Shields, R. H. Harris Jr., C. Davis, A. B. Morgan, T. E. Sutto, J. Callahan, P. C. Trulove, and H. C. De Long, *Chem. Mater.*, **2002**, 14, 2776.
2. W. H. Awad, J. W. Gilman, M. Nyden, R. H. Harris Jr., T. E. Sutto, J. Callahan, P. C. Trulove, H. C. De Long, and D. M. Fox, *Thermochim. Acta*, **2004**, 409, 3.
3. M. E. Van Valkenburg, R. L. Vaughn, M. Williams, and J. S. Wilkes, In *Proc. of the 13th Intl. Symp. on Molten Salts*, P. C. Trulove, H. C. De Long, R. A. Mantz, G. R. Stafford, and M. Matsunaga, Eds., The Electrochemical Society, Pennington, NJ, 2002, 112.
4. H. L. Ngo, K. LeCompte, L. Hargens, and A. B. McEwen, *Thermochim. Acta*, **2000**, 357, 97.

5. J. G. Huddleston, A. E. Visser, W. M. Reichert, H. D. Willauer, G. A. Broker, and R. D. Rogers, *Green Chem.*, **2001**, 3, 156.
6. D. M. Fox, W. H. Awad, J. W. Gilman, P. H. Maupin, H. C. De Long, and P. C. Trulove, *Green Chem.*, **2003**, 5, 724.
7. R. E. Lyon, "Plastics and Elastomers", Chapter 3, in *Handbook of Materials in Fire Protection*, C. A. Harper, Ed., McGraw-Hill, NY, 2003.
8. S. Bourbigot, J. W. Gilman, and C. A. Wilkie, *Polym. Degrad. Stab.*, **2004**, 84, 483.
9. M. E. Brown, *Introduction to Thermal Analysis, 2nd ed.*, Kluwer Academic Publishers, Dordrecht, Netherlands, 2001.
10. T. Ozawa, *Bull. Chem. Soc. Japan*, **1965**, 38, 1881.
11. J. H. Flynn and L. A. Wall, *J. Res. Nat. Bur. Stand.*, **1966**, 77A, 487.
12. H. Friedman, *J. Polym. Sci.*, **1965**, 50, 183.
13. S. Vyazovkin, *Int. Rev. Phys. Chem.*, **2000**, 19, 45.
14. J. H. Flynn, *J. Thermal Anal.*, **1983**, 27, 95.
15. A. W. Coats and J. P. Redfern, *Nature*, **1964**, 201, 68.
16. A. K. Galwey, *Thermochim. Acta*, **2003**, 399, 1.

Chapter 16

Interactions of Gases with Ionic Liquids: Experimental Approach

M. F. Costa Gomes, P. Husson, J. Jacquemin, and V. Majer

Laboratoire de Thermodynamique des Solutions et des Polymères, UMR 6003 CNRS/Université Blaise Pascal, Clermont-Ferrand, 24 avenue des Landais, 63177 Aubière Cedex, France

The interactions between gaseous molecules and room temperature ionic liquids can be investigated by determining experimentally gas solubilities as a function of temperature. Original experimental data were obtained for carbon dioxide in 1-butyl-3-methylimidazolium hexafluorophosphate and for carbon dioxide and argon in 1-butyl-3-methylimidazolium tetrafluoroborate between 30 and 70°C at atmospheric pressure and were compared with recently published values from other research groups. The experimental technique used is based on an isochoric saturation method and involves the equilibration of known amounts of dry gas and degassed solvent. The quantity of dissolved gas is calculated from the measurement of the equilibrium pressure of the saturated solution at each temperature. From the variation of the solubility with temperature, partial molar quantities of solvation such as free energy and enthalpy are derived, providing an insight on the molecular mechanisms involved in the dissolution process.

Introduction

Gas solubility measurements constitute an important source of information about the properties and structure of the solutions *(1)*. Precise determinations, covering sufficiently large temperature ranges, are often the only way to access the changes of thermodynamic properties associated with dissolution *(2)*. This is specially relevant for the case of dilute solutes for which calorimetric or volumetric determinations are very difficult.

Henry's law constants calculated from solubility data can be exactly converted to the Gibbs energies of solvation *(3)* corresponding to the change in partial molar Gibbs energy when the solute is transferred, at constant temperature, from the pure perfect gas at standard pressure to the infinitely dilute state in the solvent. For the case of gaseous solutes at low pressure this free energy of solvation can be regarded as a reasonable approximation for the Gibbs energy of solution *(1)*. Furthermore, this quantity is accessible by models of statistical thermodynamics and can be directly calculated by molecular simulation provided a realistic intermolecular potential is used. These thermodynamic functions, obtained from accurate experimental data, are thus important for developing and validating potentials characterizing solute-solvent interactions.

Besides their fundamental interest, gas solubilities in ionic liquids are also of practical importance. These data are largely used for the calculation of vapor-liquid equilibria in systems of potential technological interest namely in solvents for reaction systems *(4)* or for the development of new separation processes *(5)*. As it is an impossible task to study experimentally all the ionic liquids with a potential interest in practical applications, due to the numerous possibilities of variation of the cations and anions, it is the aim of the present work to provide insights into the facts characterizing the solvation properties of gaseous solutes.

Gas solubility was measured in two common used and commercially available ionic liquids: 1-butyl-3-methylimidazolium tetrafluoroborate, [bmim][BF$_4$], and 1-butyl-3-methylimidazolium hexafluorophosphate, [bmim][PF$_6$]. The choice of the gaseous solutes presented here reflects the two aspects refered above: carbon dioxide is frequently used in industrial mixtures, argon is a simple nonpolar gas.

The solubility of carbon dioxide is compared in the two ionic liquids and with that of argon in [bmim][BF$_4$]. The values for CO$_2$ in [bmim][PF$_6$] are checked against those published by other authors who have studied this system as a function of temperature, using different solvent samples and distinct experimental techniques. A direct comparison of the solubilities, Henry's law constants and also of the derived thermodynamic properties of solvation is possible.

Experimental

Materials

The gases used were obtained from AGA/Linde Gaz with mole fraction purities of 99.995% and 99.997% for carbon dioxide and argon, respectively. Both gases were used without further purification.

The [bmim][BF₄] was purchased from Sigma Aldrich with a stated minimum mole fraction purity of 97% and the [bmim][PF₆] was obtained from Acros Organics with a mole fraction purity greater than 97.5%. Both liquids were used without further purification but were conditioned before each solubility measurement: dried and degassed under vacuum at 70°C during 24 h. The water content of each sample was determined by Karl-Fisher titration and it was found to be 8500 ppm and 390 ppm before conditioning and 700 ppm and 112 ppm after conditioning for [bmim][BF₄] and [bmim][PF₆], respectively.

Apparatus and Operation

The solubility measurements involve the equilibration of known amounts of dry gas and degassed solvent at constant volume and the determination of the equilibrium pressure for the saturated solution maintained at constant temperature. The experimental apparatus used was described in a previous publication (6) and is represented in Figure 1.

The equilibrium cell, EC, the pressure transducer (precision manometer Druck RPT 200S, 10 to 1800 mbar, precision of 0.01% full scale), M, and the calibrated gas bulb, GB, constitute the equilibration section of the apparatus limited by stopcock 2. This part is maintained at constant temperature inside a liquid bath controlled to within $\pm 0.02°$ by means of a PID temperature controller. The temperature is measured with a platinium resistence thermometer (Hart Scientific HB-5613).

The sequence of operations followed during the gas solubility experiments is schematically represented in Figure 2.

The measurement starts with the introduction of a known quantity of solute in the gas bulb. The exact amount of gas is determined by measuring its pressure in the manometer M at constant temperature, correcting for gas imperfection. The volume of the gas bulb was previously calibrated with mercury (GB = 7.117 ± 0.008 cm^3 or 120.6 ± 0.4 cm^3 at 30°C, depending on the solubility range). An accurately known amount of ionic liquid, which varies from 3 to 5 ml and is determined by weighting, is then introduced in the equilibrium cell. After degassing and drying the pure solvent, stopcock 2 is closed and stopcock 1 is opened promoting the contact between the gas and the solvent. The volume in the equilibrium section has been previously calibrated by gas expansions.

The equilibration process is initiated by turning the stirring on. Pressure and temperature during the dissolution process are recorded in a computer via the serial interface until thermodynamic equilibrium is reached – pressure varies less than 1 Pa. Equilibrium is typically attained after 4 h.

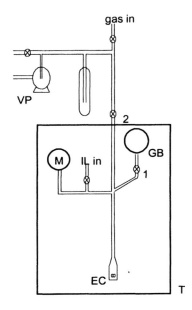

Figure 1. Isochoric solubility apparatus: VP, vacuum pump; GB, gas bulb; M, precise manometer; EC, equilibrium cell; T, thermostated bath.

As can be observed in Figure 2, the measurement of solubility at different temperatures was done by simply changing the liquid thermostat set point and waiting for a new thermodynamic equilibrium to be attained. It is thus possible to make measurements over large temperature ranges with a single loading of ionic liquid and gaseous solute. Several runs with different samples were performed to check the reproducibility of the results.

Data reduction

The mole fraction of the solute in the liquid phase can be calculated from:

$$x_2 = \frac{n_2^{liq}}{n_1^{liq} + n_2^{liq}} \tag{1}$$

where n_2^{liq} is the amount of solute dissolved in the ionic liquid, $n_1^{liq} = n_1^{tot}$ is the amount of ionic liquid (the total amount of solvent can be considered, in the case of nonvolatile solvents, equal to the quantity of solvent in the liquid phase after equilibrium).

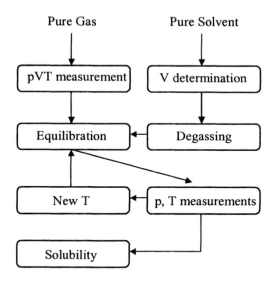

Figure 2. Block diagram of the experimental procedure used for the solubility measurements.

The quantity of solute in the liquid solution is equal to the difference between the amount of gas initially introduced in the equilibrium cell and the quantity of gas in the vapour phase in equilibrium, both quantities being determined by pVT measurements:

$$n_2^{liq} = \frac{p_i V_{GB}}{Z_2(p_i,T_i)RT} - \frac{p_{eq}(V_{tot}-V_{liq})}{Z_2(p_{eq},T_{eq})RT} \tag{2}$$

V_{GB} is the volume of the gas bulb at temperature T_i, V_{tot} the total volume of the equilibrium cell (appropriate corrections are considered for the thermal expansions in both cases) and V_{liq} the volume occupied by the liquid solution at temperature T_{eq}. Although the volume occupied by the liquid solution can be measured accurately enough in the present experimental arrangement (by gas expansions after equilibrium), the present data were obtained considering the density of the solution as equal to that of the pure solvent reported in the literature (7). p_i is the initial pressure in the gas bulb and p_{eq} the equilibrium pressure.

Henry's law constants, considered in the present case as independent of pressure, can be calculated from the mole fraction solubilities, given by eq 1, as:

$$K_H(T) \equiv \lim_{x2 \to 0} \frac{f_2(p,T,x_2)}{x_2} \cong \frac{\phi_2(p_{eq},T_{eq})p_{eq}}{x_2} \tag{3}$$

where f_2 is the fugacity of the solute and ϕ_2 its fugacity coefficient calculated by the usual way (3).

The exact expression for the standard Gibbs energy of solvation (corresponding to the Gibbs energy difference for the solute in the infinitely diluted solution and as a perfect gas) is:

$$\Delta G_2^0(T) = RT \ln\left(\frac{K_H(T)}{p^0}\right) \tag{4}$$

where p^0 is the standard pressure. Thus, from the variation with temperature of the Henry's law constants, derived thermodynamic properties pertaining to the solvation process can be determined. The enthalpy of solvation is calculated from the partial derivative of the Henry's law constants with respect to temperature at constant pressure:

$$\Delta H_2^0(T) = -T^2 \left[\frac{\partial\left(\Delta G_2^0(T)/T\right)}{\partial T} \right]_p \tag{5}$$

and similarly for the entropy of solvation:

$$\Delta S_2^0(T) = -\left[\frac{\partial\Delta G_2^0(T)}{\partial T} \right]_p \tag{6}$$

For the case of gaseous solutes at low pressures, like the systems studied in the present work, the properties of solvation calculated by eq 5 and eq 6 are reasonable approximations for the enthalpy of solution and the entropy of solutions, respectively.

Results and Discussion

Henry's Law Constants

Henry's law constants for carbon dioxide in [bmim][PF$_6$] are compared in Figure 3 with those for [bmim][BF$_4$] previously published *(6)*.

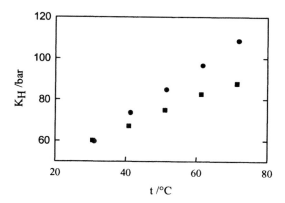

Figure 3. Henry's law constants for carbon dioxide in [bmim][BF$_4$] (6), ■, *and [bmim][PF$_6$],* ●, *as a function of temperature.*

It is observed that the Henry's law constants are slightly higher for [bmim][PF₆] at temperatures above 40°C. The solubility of carbon dioxide which is the quantity directly determined during our experiments, is similar in both solvents and decreases with temperature in the range covered. The magnitude of the solubility, expressed in mole fraction for a pressure of 101325 Pa, varies from 0.9 to 2×10^{-2} between 30 and 70°C.

For the case of CO_2 in [bmim][PF₆], comparison with literature published values is possible as this system has been studied by a number of authors using different samples and different experimental protocols although in some cases the pressure range covered was not exactly the same. Figure 4 shows Henry's law constants for this system determined by three different working groups and a deviation plot of the different results from the correlation of our data.

It is observed that, except for the case of Blanchard et al. *(8)*, the results for the Henry's law constants obtained by the different authors all lie within ±10% of our correlation. Several reasons can be evoqued to explain these deviations: the accuracy of the experimental techniques used; the need to extrapolate the experimental data obtained at high pressures to the limits of applicability of Henry's law; bad characterization of the samples used as well as presence of impurities and errors in the auxiliary properties of the ionic liquids (like the liquid density).

In Figure 5 are represented the Henry's law constants for argon and carbon dioxide in [bmim][BF₄]. The values are much larger for argon (corresponding to a much lower solubility) and their variation with temperature is opposite to that of carbon dioxide. The solubility of argon increases with temperature and, expressed in mole fraction for a partial pressure of gas equal to 101325 Pa, varies from 0.6 to 1×10^{-3} – one order of magnitude lower than for carbon dioxide.

Thermodynamic Properties of Solvation

The thermodynamic properties of solvation were calculated using eq 5 and eq 6 for the enthalpy of solvation and the entropy of solvation, respectively.

For carbon dioxide, the heats of solution are negative (exothermic process) and vary between −17 and −10 kJ mol^{-1} and −10 and −5 kJ mol^{-1} for [bmim][PF₆] and [bmim][BF₄], respectively. Entropies of solution are strongly negative and vary between −63 and −90 J mol^{-1} K^{-1} for [bmim][PF₆] and −55 and −70 J mol^{-1} K^{-1} for [bmim][BF₄].

The situation for argon is quite different as the heats of solution calculated are positive (endothermic process of dissolution) and were found to be approximately constant over the temperature range covered – average value of $+19$ kJ mol^{-1}. A much less negative entropy of solution – average value of -25 J mol^{-1} K^{-1} – was found for the dissolution of this gas in [bmim][BF$_4$].

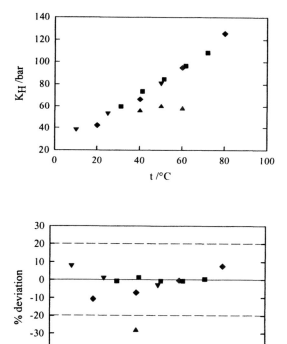

Figure 4. Comparison between the data of the literature for the Henry's law constants of carbon dioxide in [bmim][PF$_6$]: ■, this work; ▲, Blanchard et al., 2001 (8); ▼, Anthony et al., 2002 (9); ◆, Kamps et al., 2003 (10).

The values for the thermodynamic properties of solution could be compared to the ones reported in references (9) and (10) for carbon dioxide in [bmim][PF$_6$]. The values reported for the enthalpy of solution: -16.1 kJ mol^{-1} (9) and -17.24 kJ mol^{-1} (10) and for the entropy of solution: -53.2 J mol^{-1} K^{-1} (9) and -79.5 J mol^{-1} K^{-1} (10) are in accord with the ones obtained in this work.

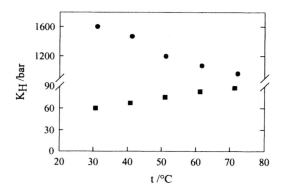

Figure 5. Henry's law constants for argon, ●, and carbon dioxide (6), ■, in [bmim][BF$_4$] as a function of temperature.

The fact that the heats of solution have different signs for the two gases studied is not surprising. A similar behavior has been found in previous studies on the solubility of these gases in molten salts like NaNO$_3$ (11), LiNO$_3$, RbNO$_3$ and AgNO$_3$ (12) probably corresponding to similar molecular mechanisms of dissolution. In the light of the empirical models used to explain the solubility of gases in molten salts (11,12) the heats of solution correspond to a balance between the contribution due to the creation of a sufficiently large cavity in the liquid solvent (endothermic process) and the ion-molecule interactions (normally an exothermic process). So it is normaly considered that negative heats of solution are observed when strong interactions occur between the solute molecules and the ions in the melt – this is probably the case for carbon dioxide but not for argon where the heat of dissolution strongly depends on the free energy of cavity formation.

Conclusion

The accuracy of gas solubility measurements in ionic liquids at low pressures and as a function of temperature could be established at ±10% by comparing the results for carbon dioxide in 1-butyl-3-methylimidazolium tetrafluoroborate obtained by different authors using different samples and various experimental techniques.

Furthermore, it was proved that accurate gas solubility measurements in ionic liquids can provide important information concerning the thermodynamic properties of solution even in the case of low solubility gases. Direct comparison with calorimetric determinations is still not possible but, by analogy with the case of aqueous solutions *(13)*, it is believed that the heats and entropies of solution obtained from the variation with temperature of Henry's law constants are reliable. These values can contribute to the understanding of the molecular mechanisms controlling the solubilization process as well as the importance of the solute-solvent interactions.

References

1. Hildebrand, J.H.; Prausnitz, J.M.; Scott, R.L. *Regular and Related Solutions*; Van Nostrand Reinhold: New York, 1970; pp 111–141.
2. Wilhelm, E.; Battino, R. *Chem. Rev.* **1973**, *73*, 1-9.
3. Prausnitz, J.M.; Lichtenthaler, R.N.; Azevedo, E.G. *Molecular Thermodynamics of Fluid-Phase Equilibria*, 3rd edition; Prentice Hall: UpperSaddle River, NJ, 1999.
4. *Ionic Liquids in Synthesis;* Wasserscheid, P.; Welton, T., Eds.; Wiley-VCH: Weinheim, Germany, 2003; pp 174-347.
5. Brennecke, J.F.; Maginn, E.J. *AIChE J.* **2001**, *47*, 2384-2389.
6. Husson-Borg, P.; Majer, V.; Costa Gomes, M.F. *J. Chem. Eng. Data* **2003**, *48*, 480-485.
7. Seddon, K.R.; Stark, A.; Torres, M.-J. In *Clean Solvents;* Moens, L.; Abraham, M.A., Eds.; ACS Symposium Series; ACS: Washington DC, 2001;Vol. 819.
8. Blanchard, L.A.; Gu, Z.; Brennecke, J.F. *J. Phys. Chem. B* **2001**, *105*, 2437-2444.
9. Anthony, J.L. ; Maginn, E.J.; Brennecke, J.F. *J. Phys. Chem. B* **2001**, *105*, 10942-10949.

10. Kamps, A. P.-S. ; Tuma, D. ; Xia, J.; Maurer, G. *J. Chem. Eng. Data* **2003**, *48*, 746-749.
11. Field, P.E.; Green, W.J. *J. Phys. Chem.* **1971**, *75*, 821-825.
12. Cleaver, B.; Mather, D.E. *Trans. Faraday Soc.* **1970**, *66*, 2469-2482.
13. Olofsson, G.; Oshodj, A.A.; Qvarnström, E.; Wadsö, I. *J. Chem. Thermodynamics* **1984**, *16*, 1041-1052.

Chapter 17

Refractive Indices of Ionic Liquids

Maggel Deetlefs[1], Michael Shara[2], and Kenneth R. Seddon[1]

[1]The QUILL Centre, Stranmillis Road, The Queen's University of Belfast, Belfast, BT9 5AG, Northern Ireland, United Kingdom
[2]Division of Physical Sciences, American Mueseum of Natural History, Rose Center, Central Park West and 79th Street, New York, NY 10024

At present, optical microscopy studies of minerals, especially diamonds, are hampered by the lack of available high refractive index (> 1.8) immersion fluids. We report here the syntheses and refractive indices of some 1-alkyl-3-methylimidazolium based ionic liquids containing polyhalide anions, which exhibit refractive indices between 1.6 and 2.23, and thus significantly extend the range of minerals which can be studied.

Introduction

Optical microscopy studies of high refractive index (RI) compounds, which themselves require immersion in high RI liquids, are currently hampered by the fact that available high RI immersion materials are either solid at room temperature, toxic, or both. For example, tin(IV) and arsenic(III) iodides with respective RIs of 2.106 and 2.23,[1] melt above 140 °C and are highly toxic. Diiodomethane saturated with sulphur,[2] is a commercially available room temperature liquid with a high RI (1.78), but this is also a very noxious mixture. It is thus apparent that there is a need for low-melting and less harmful, high RI immersion liquids, to assist in the optical examination of high RI, solid compounds *e.g.* gems.

The refractive indices of gems, diamonds in particular, are of great interest to earth scientists, since diamonds represent the deepest samples available of the earth's interior. Diamonds are formed by crystallisation in the earth's mantle at depths of 150 km or more.[3] Since diamonds are formed by crystallisation, they often contain inclusions of other minerals, which provide earth scientists with important clues to the pressure, temperature and chemical conditions in which the diamonds grew. However, the refractive index of diamond is 2.42, making the optical examination of specimens that do not have natural flat facets, *i.e.* most diamonds, very difficult. Many such diamond specimens look like frosted/etched glass, because of tiny light scattering centres on their surfaces and polishing them is expensive and time-consuming. In addition, since most commercially available refractometers are only capable of measuring refractive indices of up to 1.8,[4] the refractive index of several such diamond specimens have been examined by microscope techniques.[5] However, since available high RI compounds are solid, or toxic or both, as well as damaging to microscope equipment, the aim of the current investigation was to prepare a range of 1-alkyl-3-methylimidazolium-based, [C_nH_{2n+1}mim], ionic liquids with RIs of up to 2.4, to aid in the examination of minerals with high RIs. The negligible vapour pressure of ionic liquids, as well as the possibility of tuning their physical properties *via* the cation and anion, made them ideal candidates as less noxious alternatives to currently available high RI immersion fluids.

Interest in ionic liquids (ILs) research continues to increase rapidly, as evidenced by the exponential growth trend observed in the literature during recent years (Figure 1). Although an extensive number of topics have been investigated to date,[6] the vast majority of studies have focused on the syntheses of new ionic liquids and their use as reaction media for synthesis[7] and catalysis.[8] A much neglected topic is the physico-chemical property studies of ionic liquids.[9] Ionic liquids have been described as 'designer solvents',[10] since the choice of cation, anion, or both, potentially allows for the tailor-making of ionic liquids with pre-selected characteristics (*e.g.* moisture stability, viscosity, density, miscibility with other co-solvents).[11,12] However, one of the greatest challenges currently facing ionic liquid chemists is to construct a physico-chemical database that can be interrogated to obtain the necessary information to prepare ionic liquids with desired physical properties.[13]

In order to contribute to such a potential physico-chemical database, as well as to prepare much needed high RI liquids, we present here the preparation, characterisation and refractive index measurements of a range of 1-alkyl-3-methylimidazolium-based ionic liquids, containing a variety of polyhalide anions. We further demonstrate how systematic changes of the 1-alkyl chain-length on the imidazolium cation, as well as changes in the anion, affect the refractive indices of the salts, allowing for the tailor-making of ionic liquids with desired refractive indices

Refractive Index Measurement Methods

The refractive index of a compound is defined as the ratio of the speed of light in a vacuum to that in a medium.[14] This phenomenon is described by Snell's law (Figure 2),[15] which expresses the relationship between the refractive

Ionic Liquid Publications

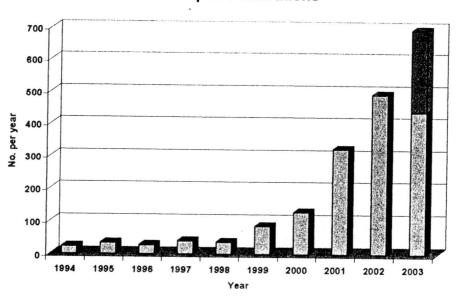

Figure 1. Ionic liquid literature growth: 1995-2003.

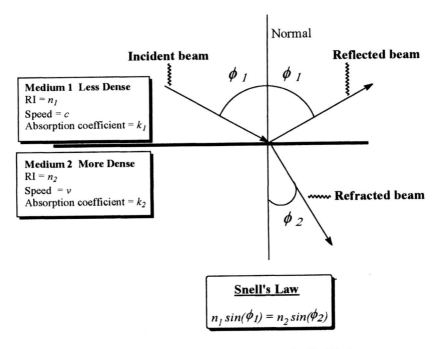

Figure 2. Refraction of light described by Snell's law.

index, the incident and transmitted angles of the light. In general, refractive indices of compounds increase with increasing atomic number of the constituent atoms[16] and furthermore, the refractive index of a compound is a physical property that can potentially be used, like a melting point, to establish its identity.

The refractive indices of liquids are generally measured using commercially available refractometers. Although such apparati provide simple and rapid measurements of RI, they can only measure refractive indices of up to *ca.* 1.8. Some alternative technologies that are available to examine higher RI liquids are reflectometry,[17] ellipsometry,[18] and optical microscopy. Although reflectometers and ellipsometers are designed to measure film thickness by measuring a film's reflectance, *R*, the refractive indices of compounds can also be determined using these instruments. In reflectometry, spectral reflectance measures the amount of light reflected by a thin film over a range of wavelengths with the incident light normal to the sample surface (compare Figure 2). Ellipsometry is similar, except that it measures reflectance at non-normal incidence and at two different polarisations. Optical constants (*n* and *k*) describe how light propagates through a film. In simple terms, the electromagnetic field (*EMF*) that describes light travelling through a material at a fixed time is given by eq. (1), where x is distance, λ is the wavelength of light, and n and k are the film's refractive index and extinction coefficient, respectively.

$$EMF = A\cos(n\frac{2\pi}{\lambda}x)\exp(-k\frac{2\pi}{\lambda}x)$$ (1)

Single interface reflection occurs when light crosses the interface between different materials. The fraction of light that is reflected by an interface is determined by the discontinuity in n and k. Reflectance of light off a material in air is given by eq. (2).

$$R = \frac{(n-1)^2 + k^2}{(n+1)^2 + k^2}$$ (2)

To see how spectral reflectance can be used to measure optical constants, consider the simple case of light reflected by a single non-absorbent material (k=0), eq. (3).

$$R = \left|\frac{n-1}{n+1}\right|^2$$ (3)

Therefore, n of the material can be determined from a measurement of R. In real materials, n varies with wavelength (since real materials exhibit dispersion) and thus RI measurements are performed at a fixed wavelength. In reflectometry measurements the reflectance of a given material is known at many wavelengths, and therefore, n at each of these wavelengths is also known. Although reflectance and ellipsometry can in theory be used to determine the RI of a liquid, in practice, the RI measurements that were obtained for a range of ionic liquids using a Filmetrics reflectometer were significantly different from the values obtained using optical microscopy. This is most probably due to the opaque nature of some of the ionic liquids studied in the current investigation.

Unfortunately, the high cost of reflectometers and ellipsometers render them unsuitable to perform routine RI measurements, especially since optical microscopy offers a much cheaper alternative for performing the measurements. Therefore, the refractive indices of liquids with RIs > 1.8 are often determined using optical microscopy. Optical microscopy RI measurements of liquids can be performed by immersing beads of known RI in a liquid of unknown RI. When the bead of known RI becomes 'invisible' under darkfield microscope illumination, a refractive index match has occurred. The converse of this method can also be employed *i.e.* using a gem with unknown RI and a liquid of known RI. However, using optical microscopy to study the RIs of gems or

224

other minerals suffers the disadvantages of requiring high RI, transparent liquids.

Yet another technique that is available to measure the RIs of liquids, and which was also used to measure the RIs of ionic liquids in the present study, is interferometry.[19] The sensor of a Fabry-Perot interferometer (Figure 3) consists of two closely spaced, partially silvered surfaces with spacing length, d. When light reaches the silvered surfaces, part of it is transmitted each time the light reaches the second surface, resulting in multiple offset beams. The large number of interfering rays produces an interferogram with extremely high resolution.

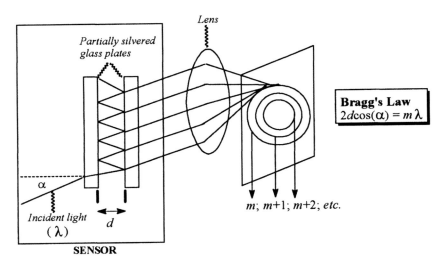

Figure 3. Fabry-Perot Interferometer Schematic.

When the sensor of the interferometer is immersed in a liquid with a RI of n, the effective value of d becomes d/n. A similar situation occurs when examining an object in water: the object appears 1.33 times closer than it actually is, since n of water is 1.33. For Bragg's law,[20] $2d\cos(\alpha) = m\lambda$ (α = angle of incidence of light on the cavity, m = integer at successive interference maxima and λ = wavelength of light used), to continue to hold true, the value of m must change. The new value of m is determined by the sensor and detector electronics. This is achieved by measuring the widths of the fringes and cross-correlating the fringe pattern with itself. In this way, the RI of any liquid, regardless of its appearance can be determined.

Preparation and Characterisation of Imidazolium-based Polyhalide Salts

The first attempts to prepare high RI ILs involved the synthesis of 1-alkyl-3-methylimidazolium-based salts containing tin(IV) in the anion (Figure 4).

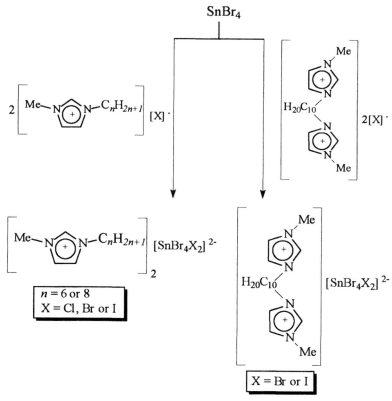

Figure 4. Syntheses of imidazolium-based hexahalostannate(IV) salts.

Tin(IV) bromide was selected as the anion precursor, since the large electron density contributions of both the metal and its halide ligands were envisaged to afford high RI salts. The cation precursors, 1-hexyl-, and 1-octyl-3-methylimidazolium chloride, bromide and iodide, were in turn selected as cations since imidazolium-based ILs containing these 1-alkyl chain lengths show melting point minima for series of ionic liquids with the same anion.[12,21]   Unfortunately,   however,   the   1-alkyl-3-methylimidazolium hexahalostannate(IV) salts exhibited melting points > 150 °C, as anticipated by the Kapustinskii equation.[22] In an attempt to lower the melting points of the tin(IV) salts, a dicationic imidazolium salt was prepared analogously, but this salt also displayed a melting point > 150 °C, thus rendering all the synthesised tin(IV) compounds unsuitable as high RI, room temperature immersion liquids. It has previously been observed that salts with 1-alkyl-3methylimidazolium cations and metal(II) halide anions (e.g. $[NiCl_4]^{2-}$, $[CoCl_4]^{2-}$, and $[PdCl_4]^{2-})^{23,24}$ have higher melting points than 1-alkyl-3methylimidazolium salts with simple inorganic anions, e.g. Cl^-, $[PF_6]^-$, etc. We therefore decided to prepare a range of 1-alkyl-3-methylimidazolium salts with polyhalide anions (Figure 5), which were envisaged to provide the necessary high electron density to give high refractive indices as well as affording ionic liquid products with low melting points.

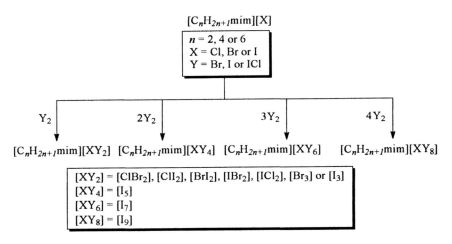

Figure 5. Syntheses of imidazolium-based polyhalide ionic liquids.

Salts containing polyhalide anions have been known for many years and their syntheses,[25] solution phase and solid state chemistry have been summarised elsewhere.[26,27] The syntheses of polyhalide salts[28] are straightforward and involve addition of halogens to quarternary halide salts, as illustrated in Figure 5. In this way, a variety of simple as well as mixed halide anions can be obtained. Recently, 1-alkyl-3-methylimidazolium trihalide ionic liquids were synthesised by Bortolini et al[29] and employed as solvents for the stereoselective iodination of alkenes and alkynes. In the present study, 1-alkyl-3-methylimidazolium salts with simple and mixed trihalide polyhalide anions were also prepared, in addition to analogous salts containing penta-, hepta-, and eneaiodide anions (Table 1).

Table I. Characterisation data for the imidazolium-based polyhalide salts.

Compound	FW / g mol^{-1}	CHN / %[a]			RI[b]	RI[c]
		C	H	N		
[C$_2$mim][IBr$_2$]	397.87	18.20 (18.11)	2.62 (2.79)	6.85 (7.04)	-	1.715
[C$_2$mim][ClI$_2$]	400.42	18.01 (17.99)	2.72 (2.77)	7.40 (7.00)	1.796	
[C$_2$mim][BrI$_2$]	444.87	16.25 (16.19)	2.47 (2.49)	6.50 (6.29)	1.833	1.833
[C$_2$mim][I$_5$]	745.67	10.00 (9.66)	1.40 (1.49)	3.73 (3.76)	-	2.23
[C$_4$mim][Br$_3$]	378.95	25.04 (25.35)	3.88 (3.99)	7.18 (7.39)	1.67	1.699
[C$_4$mim][IBr$_2$]	425.92	21.61 (21.56)	3.23 (3.54)	6.13 (6.58)	-	1.701
[C$_4$mim][ClI$_2$]	428.47	22.44 (22.42)	3.17 (3.53)	6.45 (6.53)	1.774	1.775
[C$_4$mim][BrI$_2$]	472.92	20.03 (20.31)	2.80 (3.19)	5.82 (5.92)	1.810	1.805
[C$_4$mim][I$_3$]	519.94	18.66 (18.48)	2.80 (2.90)	5.55 (5.39)	1.972	1.95
[C$_4$mim][I$_5$]	773.72	12.55 (12.42)	1.80 (1.95)	3.87 (3.62)	-	2.16
[C$_4$mim][I$_7$]	1027.52	9.81 (9.35)	1.38 (1.47)	2.94 (2.73)	-	*ca.* 2.3
[C$_4$mim][I$_9$]	1281.32	8.23 (7.99)	1.07 (1.18)	2.22 (2.19)	-	*ca.* 2.4
[C$_6$mim][IBr$_2$]	453.97	24.64 (25.05)	3.64 (4.22)	5.79 (6.17)	-	1.685
[C$_6$mim][I$_3$]	547.97	21.80 (21.92)	3.13 (3.49)	5.05 (5.11)	-	1.88

[a]Theoretical values given in parentheses; [b]Measured in 2002; error +/- 0.01; [c]Measured in 2003; error +/- 0.01

All the prepared salts are liquids at room temperature and their appearances range from red (transparent) to purple (opaque). In general, as the halide content of the anion increases, the product colour deepens. ^{1}H-NMR spectroscopic data of the salts indicated that halogenation of the imidazolium ring had not occurred. The formulations of the salts were confirmed by microanalyses; a structural study will be reported elsewhere.[30]

Negative mode Electrospray Ionisation Mass Spectrometry (ESI-MS) was used to investigate the composition of the trihalide anions (Table II), which all displayed the expected anion fragment as well as additional species previously found to form due to decomposition, exchange or aggregation of the anions. Unfortunately, negative mode ESI-MS could not be employed to determine the composition of the heavier counterions, since ESI-MS analysis requires the use of a carrier solvent, usually acetonitirile, which affects the equilibria of the anions. In the solid state, polyiodide structures in particular, represent some of the most common examples of homoatomic catenation or clustering. With very few exceptions, the structural entities can be divided into the fundamental and stable structural units I^-, I_2 and $[I_3]^-$.[31] These three structural units are most often observed in the crystalline state of compounds containing polyiodide anions[32] and are composed of I^- or $[I_3]^-$ surrounded (solvated) by neutral I_2. For example, solid state structures of compounds with $[I_5]^-$ as counterion have been shown exist in the form $[I(I_2)_2]^-$ [33] Furthermore, the stability constant for iodine in acetonitrile (eq. 4) has been found to be in the order of 10^7 so that under ESI-MS experimental conditions (*viz.* sample dissolution in acetonitrile), I_2 is virtually completely consumed and only $[I_3]^-$ and excess $[I]^-$ are present.[34,35] With this in mind, it is therefore not surprising that the ESI-MS spectra of the $[I_5]^-$ compounds recorded in the present study, only exhibited signals for I^- and $[I_3]^-$.

$$I^- + I_2 \rightleftharpoons [I_3]^-$$ (4)

As already mentioned, the structural nature of the polyhalide anions are part of an ongoing investigation and future work will include their characterisation using Fast Atom Bombardment (FAB) mass spectrometry and Raman spectroscopy. FAB mass spectrometry does not require the use of a carrier solvent and should thus eliminate the solvent-induced disproportionation of the polyhalide anions, while Raman spectroscopy has previously been used to deliver clues regarding the structures of the same anions. For the purposes of the current investigation, however, our aim was simply to determine whether imidazolium-based polyhalide ionic liquids could serve as high RI fluids before performing detailed structural studies. Nevertheless, the mass spectral data of the compounds supports the expected disproportionation shown in eq. (5).

$$2[X\text{-}Y_2]^- \rightleftharpoons [X_2Y]^- + [Y_3]^-$$

X or Y = Cl, Br or I (5)

Table II. ESI mass spectral data of polyhalide ILs in negative-ion mode.[a]

Compound	Counter-ion[b]	Additional mass peaks[b]	
$[C_2mim][ClI_2]$	$[ClI_2]^-$ (289, 9)	$[I_3]^-$ (381, 100)	$[ICl_2]^-$ (197, 3)
$[C_4mim][ClI_2]$	$[ClI_2]^-$ (289, 14)	$[I_3]^-$ (381, 100)	$[ICl_2]^-$ (197, 8)
$[C_2mim][BrI_2]$	$[BrI_2]^-$ (333, 27)	$[I_3]^-$ (381, 100)	$[IBr_2]^-$ (287, 6)
$[C_4mim][BrI_2]$	$[BrI_2]^-$ (333, 38)	$[I_3]^-$ (381, 100)	$[IBr_2]^-$ (287, 5)
$[C_4mim][Br_3]$	$[Br_3]^-$ (240, 100)	$\{[C_4mim]Br_2\}^-$ (299, 6)	$\{[C_4mim]Br_3\}^-$ (379, 43)
$[C_2mim][I_3]$	$[I_3]^-$ (381, 100)	$[I]^-$ (127, 16)	-
$[C_4mim][I_3]$	$[I_3]^-$ (381, 100)	$[I]^-$ (127, 6)	-
$[C_2mim][IBr_2]$	$[IBr_2]^-$ (287, 100)	$[BrI_2]^-$ (334, 12)	$[I_3]^-$ (381, 5)
$[C_4mim][IBr_2]$	$[IBr_2]^-$ (287, 100)	$[BrI_2]^-$ (334, 4)	$[I_3]^-$ (381, 3)
$[C_6mim][IBr_2]$	$[IBr_2]^-$ (287, 100)	-	-
$[C_2mim][I_5]$	$[I_3]^-$ (381, 100)	-	-
$[C_4mim][I_5]$	$[I_3]^-$ (381, 100)	$[I]^-$ (127, 14)	-

[a]Using CH_3CN as infusion solvent; [b]m/z values refers to ^{79}Br and ^{35}Cl.

The refractive indices of polyhalide ionic liquids prepared in the current investigation are given in Table I.. Furthermore, the RIs measured in 2002 were determined using optical microscopy techniques, whereas the measurements made in 2003 were determined using a Fabry-Perot interferometer. The RI data indicate that the values remained essentially the same over a period of one year, which is encouraging for the development of these compounds as stable, high RI ionic liquids. The RI measurements also show the reliability of the data since two techniques were employed to determine the RI values and only minor discrepancies exist. Finally, and most importantly, from Table I it is evident that a range of imidazolium-based ionic liquids are available with variable RIs. It must be mentioned here that although the RIs of $[C_4mim][I_7]$ and $[C_4mim][I_9]$ were included in the RI data analysis, their values are currently being re-evaluated since they have as yet only been determined using a sensor calibrated for RI values up to 1.8. Nevertheless, the values indicate how one and two additional equivalents of I_2, respectively, generally increases the salts RIs.

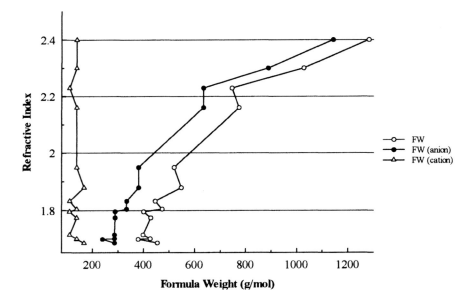

Figure 6. Plot of RIs vs. cation, anion and total FW of polyhalide ILs.

Figure 6 indicates that the RIs of the ionic liquids are governed by a primary anion and secondary cation effect. In general, the RIs of the compounds increase as the anion and total compound mass increases, provided the 1-alkyl chain-length of the cation remains short. This is better illustrated in Figure 7, which shows that for the same anion mass, 286.7 g mol^{-1}, the refractive indices of [C$_2$mim], [C$_4$mim] and [C$_6$mim][IBr$_2$] gradually increases, as the 1-alkyl chain length decreases and the total formula weight increases. The RI values of [C$_2$mim] and [C$_4$mim][I$_5$] as well as the data of [C$_2$mim] and [C$_4$mim][CII$_2$] exhibit a similar trend (see Table I), implying that the RI increase trend also occurs in other polyhalide compounds. An exciting implication of this trend is the potential to employ it as a qualitative predictive tool for the preparation of imidazolium-based polyhalide compounds with specific RI values.

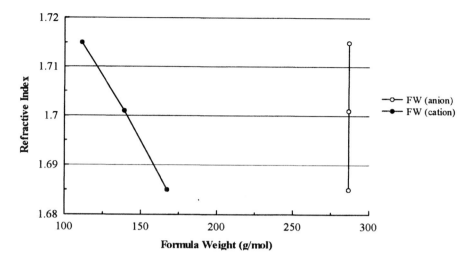

Figure 7. Plot of RI vs. *cation, anion and total FW of [C_nmim][IBr_2] ILs.*

Summary

A range of 1-alkyl-3-methylimidazolium salts containing a variety of polyhalide counter anions have been prepared and their refractive indices determined using both optical microscopy and interferometry. Using these two techniques, the reliability of the data is demonstrated since only minor discrepancies are observed for the RI values. The refractive index values of the compounds fall between *ca.* 1.6 and 2.23 and are higher for the salts containing heavy anions and lighter cations. Indeed, for the same anion, the RIs of a series increase as the 1-alkyl chain-length decreases *i.e.* a primary anion and secondary cation dependence is observed. An exciting implication of this trend is the potential to employ it as a qualitative predictive tool for the predictive preparation of imidazolium-based polyhalide compounds with specific RIs. Finally, the obtained RI data indicates that product decomposition does not occur over extended periods of time, which is encouraging for the development of these compounds as alternatives to existing, toxic, high RI fluids.

Acknowledgements

We would like to thank QUILL for the funding of this study.

References

1. Lide, D. R. *CRC Handbook of Chemistry and Physics,* 75th Ed, CRC Press, USA, **1994**, p 4-41, 4-108.
2. (a) URL: http://www.inwit.com/inwit/writings/refractiveindices.html, **2002**; (b) URL: http://www.cargille.com, **2003**.
3. URL: http://www.amnh.org/exhibitions/diamonds/formation.html, **2003**.
4. See for example:
    (a) URL: http://www.misco.com/refractometer/abbe_refractometer.html, **2001**; (b) URL: http://www.topac.com/atrsw.html, **2002**; (c) URL: http://www.leica-microsystems.com/website/lms.nsf, **2002**;
    (d) URL: http://www.rudolphresearch.com/refractometer/index.htm, **2003**.
5. Nesse, W. D. *Introduction to Optical Mineralogy,* 3rd Ed., Oxford University Press, Great Britain, **2003**, pp 480.
6. See for example: (a) Rogers, R. D.; Seddon, K. R. (eds.), *Ionic Liquids: Industrial Applications for Green Chemistry*, ACS Symp. Ser., Vol. 818 (American Chemical Society, Washington D.C., **2002**); (b) Rogers, R. D. Seddon, K. R. (eds.), *Ionic Liquids as Green Solvents: Progress and Prospects,* ACS Symp. Ser., Vol. 856 (American Chemical Society, Washington D.C., **2003**); (c) Rogers, R. D., Seddon, K. R.; Volkov, S. (eds.), *Green Industrial Applications of Ionic Liquids*, NATO Science Series II: Mathematics, Physics and Chemistry, Vol. 92 (Kluwer, Dordrecht, **2002**).
7. Wasserscheid, P.; Welton, T. *Ionic Liquids in Synthesis*, Wiley-VCH, Weinheim, Germany, **2002**.
8. (a) Wasserscheid, P.; Keim, W. *Angew. Chem. Int. Ed. Engl.* **2000**, *39*, 3772-3789; (b) Welton, T. *Chem. Rev.* **1999**, *99*, 2071-2083; (c) Holbrey, J. D.; Seddon, K. R. *Clean Prod. Proc.* **1999**, *1*, 223-236; (d) Sheldon, R. *Chem. Commun.* **2001**, 2399-2407; (e) Gordon, C. M. *Appl. Catal. A.* **2001**, *222*, 101-117; (f) Zhao, D.; Wu, M.; Kou, Y.; Min, E. *Catalysis Today* **2002**, *2654*, 1-33; (g) Olivier-Bourbigou. H.; Magna, H. *J. Mol. Catal. A.* **2002**, *164*, 1-19.
9. Seddon, K. R. *Green Chem.* **2002**, *4*, G25-G26.
10. Freemantle, M., *Chem. Eng. News*, **1998**, *76*, 32-33.

11. Seddon, K. R.; Stark, A.; Torres, M-J. in *Clean Solvents: Alternative Media for Chemical Reactions and Processing*, Eds. Abraham, M.; Moens, L. ACS Symp. Ser., Vol. 819 (American Chemical Society, Washington D.C., 2002), pp. 34-49.

12. Torres, M-J. PhD thesis, Queen's University of Belfast, UK, **2001.**

13. URL: http://www.iupac.org/projects/2003/2003-020-2-100.html, **2003.**

14. Samsonov, G. V. *Handbook of the Physicochemical Properties of the Elements*, IFI-Plenum, New York, USA, **1968.**

15. Joyce, W. B.; Joyce, A., *J. Opt. Soc. Amer.* **1976,** *1,* 1-8.

16. Sangwal, K.; Kucharczyk, W. *J. Phys. D: Appl. Phys.*, **1987,** *20,* 522-525.

17. URL: http://www.filmetrics.com, **2003.**

18. URL: http://www.jawoollam.com, **2003.**

19. URL: http://hyperphysics.phy-astr.gsu.edu/hbase/phyopt/fabry.html, **2003**.

20. Atkins, P. W. *Physical Chemistry,* 5th Ed., Oxford University Press, Great Britain, **1994,** p 764.

21. Holbrey, J. D.; Seddon, K. R. *J. Chem. Soc. Dalton. Trans.,* **1999,** 2133-2139.

22. Kapustinskii, A.F. *Q. Rev., Chem. Soc.*, **1956,** *10,* 283-300.

23. Hitchcock, P. B.; Seddon, K. R.; Welton, T. *J. Chem. Soc. Dalton. Trans.* **1993,** 2639-2641.

24. Hardacre, C.; Holbrey, J. D.; McCormac, P. B.; McMath, S. E. J.; Nieuwenhuyzen, M.; Seddon, K. R. *J. Mater. Chem.* **2001,** *11,* 346-350.

25. Moeller, T. *Inorganic Syntheses,* McGraw-Hill, USA, **1957,** Vol. V, p 167-178.

26. Popov, A. I., *Halogen Chemistry,* Gutmann, V. Ed., Academic Press, NY, **1967,** Vol. I, p 225.

27. Downs, A. J.; Adams, C. J. *Comprehensive Inorganic Chemistry,* Bailar, J. C.; Emeleus, H. J.; Nyholm, R.S.; Trotman-Dickenson, Eds., Pergamon, Oxford, **1973,** Vol. II, p. 1534 *et seq.*

28. Hope, E. G. *Annu. Rep. Prog. Chem. Sect. A.,* **2000,** *96,* 135-146.

29. Bortolini, O.; Bottai, M.; Chiappe, C.; Conte, V.; Pieraccini, D. *Green Chem.*, **2002,** *4,* 621-627.

30. Berg, R. W.; Deetlefs, M.; Shara, M.; Seddon, K. R. *unpublished results.*

31. Svensson, P. H.; Kloo, L. *J. Chem. Soc. Dalton Trans.* **2000,** 2449-2455.

32. Greenwood, N. N.; Earnshaw, A. *Chemsitry of the Elements,* 2nd Ed., **1997,** Pergamon, Oxford, p 835.

33. Hach, R. J.; Rundle, R. E. *J. Am. Chem. Soc.*, **1951,** *73,* 4321-4325.

34. Desbarres, J. *Bull. Soc. Fr.* **1961,** *58,* 955-957

35. Nelson, I. V.; Iwamoto, R. T. *J. Electroanal. Chem.* **1964,** *7,* 218-220.

Chapter 18

Amphiphilic Self Organization in Ionic Liquids

Gary A. Baker[1] and Siddharth Pandey[2]

[1]Bioscience Division, Los Alamos National Laboratory, MS J586,
Los Alamos, NM 87545
[2]Department of Chemistry, Indian Institute of Technology, Delhi, Hauz
Khas, New Delhi 110016, India

Despite clear interest, progress toward the controlled formation of amphiphilic assemblies in ionic liquids (or their purely inorganic ancestor, the low-melting molten salt) is still in its infancy. In this chapter, we provide brief historical perspective, highlight results from the recent literature and speculate on the future of this exciting area of research.

Background

'Second-generation' ionic liquids (air and moisture stable organic-containing molten salts) have garnered global attention not least because they hold the simultaneous potential for novel, improved and/or eco-friendly chemistry. The latter point is manifest by strict adherence to the guiding principles of Green Chemistry (*1*). The first two aspects dealing with unique

234

and advanced chemistry are bolstered by both the introduction of task-specific ionic liquids (2) and the fact that ionic liquids are, in general, 'modular' solvents that allow for facile synthetic variation. That is, for any given parent structure [A⁺][B⁻] one can in principle generate a truly huge superset of possibilities simply by iterative variation in the cation system, its substitution pattern and/or anion choice. Indeed, Seddon has remarked that over 10^{18} simple salts that qualify as ionic liquids might be possible (3). For comparison, about 500 molecular organic solvents find use today; of these, a few dozen—mostly volatile organic compounds (VOCs)—comprise the bulk of industrial usage. It is worth stressing that at this juncture the number and variety of well characterized and/or commercial ionic liquids is still quite small compared with molecular organic solvents available to the chemist.

Several principal strategies have emerged toward the creation of cleaner and more sustainable chemical technologies. The ultimate goal of course is elimination of the solvent altogether (4). For many applications performing a reaction 'neat' is not possible, however. Green alternatives to a solventless approach include the use of aqueous media (5), sub- or supercritical fluids, most notably supercritical carbon dioxide, $scCO_2$ (6), fluorous phases (7), ionic liquids (8) and mixtures thereof. Using $scCO_2$ in concert with ionic liquids in particular has attracted the interest of several research groups recently (9). Taking a lesson from the ubiquitous and highly successful application of surfactants in $scCO_2$ (10), one suspects that similar utility might be found for them within ionic liquids. Surprisingly, there is very little on record concerning the use of surfactants within ionic liquids.

The story actually begins much earlier than one might think. In fact, the term 'ionic liquid' is a constantly evolving one with an eminent prehistory including the field of classical 'molten salts' (11). Molten salts are also completely ionized solvents, however, they differ implicitly in both the melting temperature and the ion motif. That is, molten salts wholly consist of inorganic ions and even their eutectic mixtures do not melt below 150 °C. Ionic liquids on the other hand contain at least one organic ion, frequently a bulky and asymmetric cation based on an imidazolium, pyrrolidinium or phosphonium backbone. It is now widely accepted that the working definition of the term "ionic liquid" is reserved for salts that liquefy near or below 100 °C, an important subset of which is the room-temperature ionic liquid (8).

In the section that follows we provide an account of research relating to the use of surfactants in ionic liquids. While the information presently available is relatively limited and progress has been slow, it has been punctuated with some significant victories. To the best of the authors' knowledge, Figure 1 provides a fairly inclusive and updated list of the surfactants studied in ionic liquids so far. Although this is a modest inventory it can be safely assumed that this will not

236

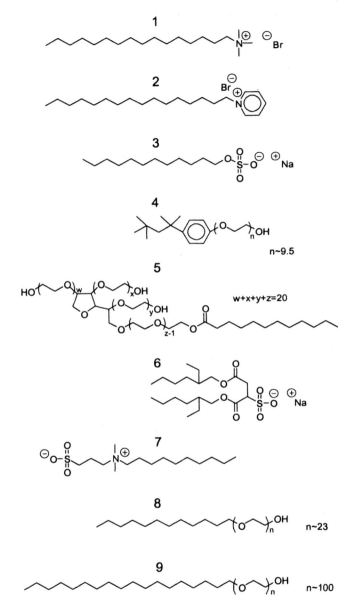

Figure 1. Structures of surfactants studied in ionic liquids. 1 CTAB; 2 HDPB; 3 SDS; 4 Triton X-100; 5 Tween 20; 6 AOT; 7 CS; 8 Brij 35; 9 Brij 700.

remain the case for long. It is our hope that this chapter serve as a *reveille* to the growing ranks of ionic liquid researchers out there.

Surfactants in Ionic Liquids: A Brief Tour

To date, reports of surfactants in molten salts or ionic liquids have been few and far between. In advance of any such account, Steigman and Shane communicated in 1965 that micelle formation was possible for long-chain fatty acids in concentrated sulfuric acid (*12*). Using surface tension and light scattering measurements these authors learned, for example, that stearic acid formed micelles in 97.3% (w/w) H_2SO_4. Although this demonstrated that micelle formation was feasible in a highly ionized solvent it should be noted that this still constitutes a considerable water level on a molar basis (~2.8 M). On the other hand, similar water levels are often associated with ionic liquids in common use. The earliest evidence for micellization of surfactants in a 'low-melting' molten salt is the work of Bloom and Reinsborough beginning a few years later in the late 1960s on pyridinium chloride melts near 150 °C (*13*).

Although ethylammonium nitrate, [EtNH$_3$][NO$_3$], an organic ionic liquid which melts around 14 °C, had been known since 1914 (this ionic liquid has the distinction of being the first one on record) (*14*) it was not until the early 1980s that Evans et al. (*15*) reported that [EtNH$_3$][NO$_3$] supported micelle formation by certain surfactants. Using classical and quasi-elastic light scattering (*15a*), hydrodynamic radii of 14 and 22 Å were estimated for tetradecylpyridinium bromide and hexadecylpyridinium bromide **2**, respectively, in [EtNH$_3$][NO$_3$]. From the change in critical micellar concentration (CMC) with surfactant chain length, the free energy of transferring a methylene group from the fused salt into the interior of a micelle was found to be –400 cal/mol compared with –680 cal/mol in water (*15c*). The fact that [EtNH$_3$][NO$_3$] is a slightly better solvent than water for hydrocarbons accounts at least in part for the 7–10 fold higher CMCs observed in [EtNH$_3$][NO$_3$]. Further, measured second virial coefficients were reasonably well described by a hard-sphere potential reflecting the effective electrostatic screening in this completely ionized solvent. Later, Chang studied a sodium dodecyl sulfate **3** + decane + ethylenediamine/ammonium/potassium nitrate molten salt (an energetic eutectic of interest to the explosives community, m.p.~104 °C) mixture using electrical conductivity (*16*). Although the results were not definitive, they did suggest a structural change from a globular droplet to a bicontinuous lamellar form as 1-pentanol was added. Although CMC values were 4 times higher than in water, direct comparison is not meaningful as these studies were conducted at 120 °C.

More recently, Friberg et al. (*17*) investigated 1-butyl-3-methylimidazolium hexafluorophosphate, [bmim][PF$_6$], solubilization in a Brij 30–water system (Brij 30 is equivalent to **8** for n=4) using small angle x-ray diffraction. Notably, the system formed a lamellar liquid crystal which solubilized large amounts of [bmim][PF$_6$] (15% w/w) without changing the dimensions of the amphiphile layer *or* the water level dependence of interlayer spacing.

In addition to its role as solvent, an appropriately designed ionic liquid may also function as a surfactant or cosurfactant. This is perhaps not too strained an assertion given the structural likeness of many existing ionic liquids to surfactants, phase transfer catalysts, electrolytes, ion exchangers and the like. A very interesting study that expands upon this concept was recently reported by Davis and co-workers (*18*). In this work, four novel ionic liquids formulated from imidazolium cations appended with long fluorous 'ponytails' were found to emulsify perfluorohexane into the 'conventional' ionic liquid 1-hexyl-3-methylimidazolium hexafluorophosphate, [C$_6$mim][PF$_6$]. While perfluorohexane and [C$_6$mim][PF$_6$] formed a biphase in the absence of surfactant *immediately* upon cessation of mechanical agitation, [C$_6$mim][PF$_6$] saturated in any of these fluorous ionic liquids formed dispersions with perfluorohexane that persisted for weeks. In a similar approach, by dispersing amide group enriched glycolipids in ether-bearing ionic liquids, authors Kimizuka and Nakashima were able to form stable, thermally labile bilayer membranes, so-called 'ionogels' (*19*).

In the last several months, three additional papers on the topic of surfactant action on or in an ionic liquid have found print. The first, a paper from the Armstrong laboratory explores the possibility of normal micelle formation for surfactants **3**, **6**, **7**, **8**, and **9** in [bmim][PF$_6$] and [bmim]Cl using surface tensiometry and inverse gas chromatography (IGC) (*20*). In line with earlier studies (*15*, *16*), their results indicate that CMCs in the ionic liquid are about an order of magnitude larger than in water. Their IGC results in concert with linear free energy relationships also show that above the CMC the hydrogen bond basicity increases dramatically for addition of the two nonionic surfactants **8** and **9** to [bmim][PF$_6$].

The preparation of simple or multiple emulsions of ionic liquids stabilized solely by the interfacial adsorption of solid silica nanoparticles is also an interesting possibility realized by Binks et al. (*21*). Although surfactants are not involved here we include this contribution because these systems offer significant advantages in the pursuit of clean chemistry.

The final topic of our discussion is a recent paper from one of the authors' labs (*22*). Using methods detailed earlier (*23*), the solvatochromic responses of two fluorescent probes–pyrene and 1,3-bis(1-pyrenyl)propane (BPP)–were examined for increasing levels of surfactants **1**, **3**, **4**, **5**, **8**, and **9** in the ionic liquid 1-ethyl-3-methylimidazolium bis(trifluoromethylsulfonyl) imide, [emim][Tf₂N]. Our interest was based on the prospect of developing a rapid and convenient optical method for estimating CMC values using miniscule ionic liquid samples. The bases for the two probe responses are a pertubation in the

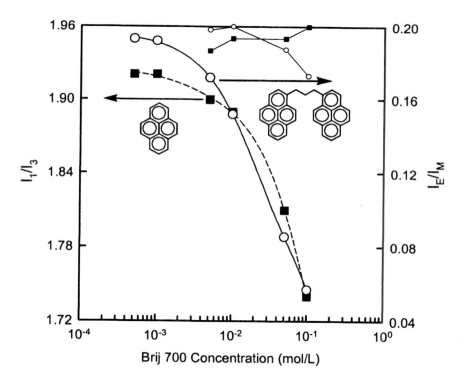

Figure 2. Effects of Brij 700 concentration on the measured pyrene I_1/I_3 (■) and BPP I_E/I_M (○) ratios in [emim][Tf₂N]. As a benchmark, results for CTAB (shown as the smaller matching symbols) are included on the same respective axes.

cybotactic region polarity [pyrene's I_1/I_3 emission ratio] or mobility [BPP's I_E/I_M ratio] at the onset of micellization. Briefly, all nonionic surfactants studied provided clear evidence for micellization; see Figure 2 for results using **9**. The right and left ordinate axes were independently scaled to highlight the fact that both probes responded similarly to changes in surfactant level lending validity to this approach for CMC determination. The sharp decrease in I_1/I_3 is consistent with a reduced polarity resulted from the pyrene residing in the hydrophobic core of a normal micelle. The decrease in I_E/I_M reflects an increased microviscosity. Both behaviors are consistent with micelle formation for known surfactants in aqueous systems at the CMC. Also, shown for comparison in Figure 2 are results for the cationic surfactant **1** for which no aggregation was observed over the concentration range studied (5×10^{-3} to 0.10 mol/L); we also note that **3** did not even dissolve in [emim][Tf$_2$N] at 'aqueous' CMC levels at room temperature. Results from both probes indicate that the CMC follows the general order **9**<<**8**<**5**≈**4**. As expected, CMC values are invariably larger than in water and range from about 10 mM [**9**] to over 50 mM [**4**], based on either method.

Future Prospectus

The incorporation of surfactants adds a new dimension to the already growing area of ionic liquids with benefits both predictable and unforeseen. One possible destination, given the identification of suitable amphiphiles, might be the formation of ionic liquid-in-scCO$_2$ microemulsions extending the applicability of supercritical fluids, for example.

Because the possibilities are essentially boundless, combinatorial discovery methods will prove invaluable. High throughput optical screening methods such as those based on platereader fluorescence polarization/anisotropy (*24*) or fluorescence correlation spectroscopy (FCS) (*25*) are attractive options because both offer rapid, high sensitivity analysis amenable to microliter volumes. Sample preparation methods involving parallel low-volume liquid handling schemes that allow for rapid array formation, for example, contact pin printing onto microscope slides (*26*), would dovetail perfectly with such tools.

In this chapter, we hope that we have stirred some interest in the area of surfactant-modified ionic liquid phases. The next decade is certain to witness many critical advances in this area with potentially profound implications for environmentally responsible separations and for green synthesis. In short, we believe the landscape looks very promising indeed!

Acknowledgements

GAB would like to thank LANL for the award of Director's and Frederick Reines Fellowships. SP acknowledges the Sandia–University Research Program for partial financial support of this work. The authors also acknowledge stimulating conversations with members of the burgeoning ionic liquid community at Los Alamos National Laboratory, particularly Sheila N. Baker, T. Mark McCleskey, Anthony K. Burrell, Bill Tumas and Warren J. Oldham.

References

1. Anastas, P. T.; Warner, J. C. *Green Chemistry: Theory and Practice*; Oxford University Press: New York, 1998; p 30.
2. (a) Visser, A. E.; Swatloski, R. P.; Reichert, W. M.; Mayton, R. D.; Sheff, S.; Wierzbicki, A.; Davis, J. H., Jr.; Rogers, R. D. *Environ. Sci. Technol.* **2002**, *36*, 2523. (b) Cole, A. C.; Jensen, J. L.; Ntai, I.; Tran, K. L. T.; Weaver, K. J.; Forbes, D. C.; Davis, J. H., Jr. *J. Am. Chem. Soc.* **2002**, *124*, 5962.
3. Seddon, K. R. In *The International George Papatheodorou Symposium: Proceedings*; Boghosian, S., Dracopoulos, V., Kontoyannis, C. G., Voyiatzis, G. A., Eds.; Institute of Chemical Engineering and High Temperature Chemical Processes: Patras, Greece, 1999; pp 131-135.
4. Tanaka, K. *Solvent-Free Organic Synthesis*; Wiley-VCH: New York, 2003.
5. Li, C.-J. ; Chan, T.-H. *Organic Reactions in Aqueous Media*; John Wiley & Sons: New York, 1997.
6. (a) *Chemical Synthesis using Supercritical Fluids*; Jessop, P. G., Leitner, W., Eds.; Wiley-VCH: Weinheim, 1999. (b) Eckert, C. A.; Knutson, B. L.; Debenedetti, P. G. *Nature* **1996**, *383*, 313, and references cited therein.
7. (a) A review that offers practical advice on fluorous techniques has been published. Barthel-Rosa, L. P.; Gladysz, J. A. *Coord. Chem. Rev.* **1999**, *192*, 587. (b) Horvath, I. T. *Acc. Chem. Res.* **1998**, *31*, 641. (c) Studer, A.; Hadida, S.; Ferritto, R.; Kim, S. Y.; Jeger, P.; Wipf, P.; Curran, D. P. *Science* **1997**, *275*, 823.
8. (a) *Ionic Liquids in Synthesis*; Wasserscheid, P., Welton, T., Eds.; Wiley-VCH: Weinheim, 2003. (b) *Ionic Liquids as Green Solvents: Progress and Prospects*; Rogers, R. D.; Seddon, K. R., Eds.; ACS Symposium Series 856, American Chemical Society: Washington, DC, 2003.
9. (a) Lu, J.; Liotta, C. L.; Eckert, C. A. *J. Phys. Chem. A* **2003**, *107*, 3995. (b) Baker, S. N.; Baker, G. A.; Kane, M. A.; Bright, F. V. *J. Phys. Chem. B* **2001**, *105*, 9663. (c) Laszlo, J. A.; Compton, D. L. *Biotechnol. Bioeng.*

2001, *75*, 181. (d) Kazarian, S. G.; Briscoe, R. J.; Welton, T. *Chem. Commun.* **2000**, 2047. (e) Blanchard, L. A.; Hancu, D.; Beckman, E. J.; Brennecke, J. F. *Nature* **1999**, *399*, 28.

10. (a) Sarbu, T.; Styranec, T.; Beckman, E. J. *Nature* **2000**, *405*, 165. (b) Kane, M. A.; Baker, G. A.; Pandey, S.; Bright, F. V. *Langmuir* **2000**, *16*, 4901. (c) Johnston, K. P.; Harrison, K. L.; Clarke, M. J.; Howdle, S. M.; Heitz, M. P.; Bright, F. V.; Carlier, C.; Randolph, T. W. *Science* **1996**, *271*, 624.

11. For an excellent historical account see: Wilkes, J. S. *Green Chem.* **2002**, *4*, 73.

12. Steigman, J.; Shane, N. *J. Phys. Chem.* **1965**, *69*, 968.

13. (a) Reinsborough, V. C. *Aust. J. Chem.* **1970**, *23*, 1473. (b) Bloom, H.; Reinsborough, V. C. *Aust. J. Chem.* **1969**, *22*, 519. (c) Reinsborough, V. C.; Valleau, J. P. *Aust. J. Chem.* **1968**, *21*, 2905. (d) Bloom, H.; Reinsborough, V. C. *Aust. J. Chem.* **1967**, *20*, 2583. (e) Reinsborough, V. C.; Bloom, H. *Aust. J. Chem.* **1968**, *21*, 1525.

14. Walden, P. *Izv. Imp. Acad. Nauk.* **1914**, 405.

15. (a) Evans, D. F.; Yamauchi, A.; Wei, G. J.; Bloomfield, V. A. *J. Phys. Chem.* **1983**, *87*, 3537. (b) Evans, D. F.; Kaler, E. W.; Benton, W. *J. Phys. Chem.* **1983**, *87*, 533. (c) Evans, D. F.; Yamauchi, A.; Roman, R.; Casassa, E. Z. *J. Colloid Interface Sci.* **1982**, *88*, 89.

16. Chang, D. R. *Langmuir* **1990**, *6*, 1132.

17. Friberg, S. E.; Yin, Q.; Pavel, F.; Mackay, R. A.; Holbrey, J. D.; Seddon, K. R.; Aikens, P. A. *J. Dispersion Sci. Technol.* **2000**, *21*, 185.

18. Merrigan, T. L.; Bates, E. D.; Dorman, S. C.; Davis, J. H., Jr. *Chem. Commun.* **2000**, 2051.

19. Kimizuka, N.; Nakashima, T. *Langmuir* **2001**, *17*, 6759.

20. Anderson, J. L.; Pino, V.; Hagberg, E. C.; Sheares, V. V.; Armstrong, D. W. *Chem. Commun.* **2003**, 2444.

21. Binks, B. P.; Dyab, A. K. F.; Fletcher, P. D. I. *Chem. Commun.* **2003**, 2540.

22. Fletcher, A.; Pandey, S. *Langmuir* **2003**, *20*, 33.

23. (a) Pandey, S.; Fletcher, K. A.; Baker, S. N.; Baker, G. A. *Analyst* **2004**, *129*, 569. (b) Baker, S. N.; McCleskey, T. M.; Pandey, S.; Baker, G. A. *Chem. Commun.* **2004**, 940. (c) Fletcher, K. A.; Pandey, S. *J. Phys. Chem. B* **2003**, *107*, 13532. (d) Fletcher, K. A.; Baker, S. N.; Baker, G. A.; Pandey, S. *New J. Chem.* **2003**, *27*, 1706. (e) Baker, G. A.; Baker, S. N.; McCleskey, T. M. *Chem. Commun.* **2003**, 2932. (f) Baker, S. N.; Baker, G. A.; Bright, F. V. *Green Chem.* **2002**, *4*, 165. (g) Fletcher, K. A.; Pandey, S. *Appl. Spectrosc.* **2002**, *56*, 266. (h) Fletcher, K. A.; Pandey, S. *Appl.*

Spectrosc. **2002**, *56*, 1498. (i) Fletcher, K. A.; Storey, I. A.; Hendricks, A. E.; Pandey, S.; Pandey, S. *Green Chem.* **2001**, *3*, 210.

24. (a) Baker, G. A.; Pandey, S.; Bright, F. V. *Anal. Chem.* **2000**, *72*, 5748. (b) Baker, G. A.; Betts, T. A.; Pandey, S. *J. Chem. Educ.* **2001**, *78*, 1100.

25. Werner, J. H.; Baker, S. N.; Baker, G. A. *Analyst* **2003**, *128*, 786.

26. Cho, E. J.; Bright, F. V. *Anal. Chem.* **2002**, *74*, 1462.

Chapter 19

Surface Chemistry and Tribological Behavior of Ionic Liquid Boundary Lubrication Additives in Water

Benjamin S. Phillips[1], Robert A. Mantz[2], Paul C. Trulove[3], and Jeffrey S. Zabinski[2,*]

[1]Universal Technology Corporation, 1270 North Fairfield Road, Dayton, OH 45432–2600
[2]U.S. Air Force Research Laboratories, Materials and Manufacturing Directorate, Nonmetallic Materials Division, Nonstructural Materials Branch, Wright-Patterson Air Force Base, OH 45433–7750
[3]AFOSR/NL, 4015 Wilson Boulevard, Arlington, VA 22203

Ionic liquids have been intensely studied regarding their properties as an environmentally friendly solvent. Conversely, little work in the field of tribology has been accomplished to date. Ionic liquids posses many favorable properties that would suggest that they have significant potential for use in the tribology field. One area that is discussed here is the use of ionic liquids as a boundary lubricant additive for water. The chemical and tribochemical reactions that govern their behavior were evaluated for two ionic liquids. Under water lubricated conditions, silicon nitride ceramics exhibit a characteristic running in period of high friction, through which surface smoothing and tribochemical reactions lead to low friction coefficients. The running-in period of high friction is the period where the majority of the wear occurs. The use of a suitable boundary lubricant to reduce the running-in period could be of significant importance. A promising candidate for this application is ionic liquids. We present the affects that ionic liquids have on the friction and wear properties of Si_3N_4, in particular the affects on the running-in period. Tribological properties were evaluated using a pin-on-disk and reciprocating tribometers. Solutions containing 2 wt % ionic liquids were produced for testing purposes. Chemical analysis of the sliding surfaces was accomplished with x-ray photoelectron spectroscopy (XPS) and Fourier Transform Infrared Spectroscopy (FTIR). The test specimens were 1" diameter Si_3N_4 disks sliding against ¼" Si_3N_4 balls. The use of ionic liquids as a boundary lubricant additive for water resulted in dramatically reduced running-in periods for silicon nitride from thousands to the hundreds of cycles. Proposed mechanisms controlling the friction and wear include the formation of a transfer film and the inception of an electric double layer.

Introduction

The push to introduce cleaner technologies has led to the search for replacements for dangerous solvents. One promising solvent replacement is ionic liquids, which have received much attention in literature on this topic. They have many unique properties such as negligible volatility, non-flammability, high thermal stability, and a low melting point, which qualifies them as excellent choices for solvents in many synthesis reactions (1-3).

While interest in ionic liquids as environmentally friendly solvents has increased, little work has been accomplished in the field of tribology. Ionic liquid are candidates for use as lubricants due to the same properties that make them excellent solvents. Recent studies have shown potential for imidazolium based ionic liquids as lubricants (4,5). Ye et al used a ball-on-disk configuration with reciprocating motion for study of two pure imidazolium based ionic liquids, 1-methyl-3-hexylimidazolium tetraflouroborate and 1-ethyl-3-hexylimidazolium tetraflouroborate. The friction coefficient did not rise above 0.065 for either ionic liquid tested on steel substrates. The authors suggest that the mechanism that provides lubrication is the breakdown of the tetraflouroborate ion to form fluoride, B_2O_3, and BN on the substrate surface (4). Related work using imidazoline borates as a lubricant was done be Gao et al (6). Although, these materials are not classified as ionic liquids, their chemical structures have similarities. Using these chemicals as water additives on steel substrates, friction coefficients of near 0.1 were obtained in a four-ball test machine. Reports by Phillips et al show decreased friction coefficients and running in periods for Si_3N_4 utilizing ionic liquids as water additives (7). These encouraging results should lead to more extensive work performed with ionic liquids in the tribology field.

Water Lubrication of Ceramics

Since the initial report of ultra-low friction coefficients of Si_3N_4 sliding in water by Tomizawa and Fischer, significant research has been undertaken in the study of water lubrication of ceramics (8-19). Si_3N_4 is the most commonly studied ceramic, although other ceramics such at alumina and silicon carbide have also been evaluated. The work done by Tomizawa and Fischer proposed that Si_3N_4 forms a hydrated silicon oxide layer in the presence of water and that the layer is soluble, and upon dissolution the surface becomes finely polished. The mechanism controlling the low friction coefficient was ascribed to hydrodynamic lubrication (8). Under water lubricated conditions Si_3N_4 initially exhibits high friction during a running-in period, where the surface is smoothed by dissolution and wear at asperity contacts (9). Certain sliding speeds and contact pressure enhance the tribochemical reaction that forms silicon oxide on the surface (10). Although hydrodynamic lubrication was the first mechanism proposed, other mechanisms have not been ruled out (13). For example, Xu et al suggests the behavior of silicon nitride is water is controlled by both hydrodynamic lubrication and boundary lubrication by a colloidal silica layer (11). Work done under solutions of varying pH by Mizuhara strengthens this argument (12). He reports that the dissociation rate of the silica layer is the key factor in the behavior of Si_3N_4 in water.

The behavior of Si_3N_4 in water exhibits a high friction run-in period, where the majority of the wear occurs. A significant advance would be to reduce the running-in period. This could be accomplished by the use of a boundary lubricant additive to water. Reducing the running-in process would decrease the wear of Si_3N_4 and other similar materials. Typical boundary lubrication additives have several characteristics including: high load carrying capacity, low volatility, low shear strength, high stability, and environmentally friendly. As stated above, ionic liquids provide many of these characteristics and have shown promise in recent tribology testing (4).

Experimental

Ionic Liquids

Two wt.% solutions of the ionic liquids in water were evaluated in the tribology tests. The two solutions were comprised of 1-methly-3-n-butyl-imidazolium hexaflourophosphate (PF$_6$) and 1-ethyl-3-methyl imidazolium tetraflouroborate (BF$_4$).

Tribo-Testing

Friction and wear testing was performed using a pin-on-disk tribometer. The tribometer utilizes a stationary pin on a rotating disk and is setup to accommodate a 1" in diameter disc and a ¼" ball. Multiple runs of self-mated samples were performed on each disc, with the test radius varying, keeping the linear speed constant at 0.120 m/s. The test apparatus is described in more detail by Phillips et al [7].

Tribological testing was also carried out using a reciprocating tribometer. The tests were run with a 5mm stroke length and a period of 6 Hz. A normal load of 200 N was applied, which correlates to 700 MPa maximum Heritian stress at the beginning of each test. Similar to the pin on disk testing the sample and pin were self-mated pairs. All tests were carried out in air in ambient conditions. The test apparatus is described in more detail by Cavdar et al [20].

Analytical Techniques

Compositional analysis was made with an XPS system operated at 8×10^{-10} torr. Binding energy positions were calibrated using the Au 4f$_{7/2}$ peak at 83.93 eV, and the Cu 3s and Cu 2p$_{3/2}$ peaks at 122.39 and 932.47. A nominal spot size of 300um was used for analysis. Charge neutralization was accomplished by flooding the sample surface with low energy electrons.

Results

Chemical and Surface Analysis

Surface roughness and wear track areas were obtained from a contact profilometer with a resolution of 0.5 Angstroms. The results were similar to water alone where smoothing of the asperity tips in the contact area occurred for both samples. For the water only test, the maximum depth of the wear scar was approximately 6000 Angstroms with a corresponding specific wear rate of 1.5×10^{-5} mm^3/N*m. The maximum depth of the water plus ionic liquids test was approximately 2000 Angstroms with a specific wear rate of 4.58×10^{-6}

mm³/N*m. The wear rate for the water plus ionic liquid test was an order of magnitude smaller than those for pure water. The wear rates were calculated over the first 10,000 cycles, which includes the high wear running-in period. Subsequent sliding would produce little wear and therefore increased sliding distance would reduce the specific wear rates.

All composition data was taken from samples run in the oscillating friction tester. This was due to the very narrow wear scar created during the pin-on-disk test, which made it difficult to determine compositional changes with XPS (300 um spot). The wear scar generated from the oscillating friction tests was 5 x 5 mm wear scar, which enabled a more detailed study of the chemical changes within the system.

Chemical analysis on the surface was completed after the water was allowed to evaporate from the samples. XPS and other chemical analyses were performed inside and outside of the wear area as well as before and after a methanol rinse. Little difference in the composition between the inside and outside of the wear scar during XPS analysis. The differences remained small after rinsing the sample with methanol.

PF_6 tests resulted in a significant amount of fluorine, near 20% F, and small amounts of phosphorus, <5%, present on the surface. After a methanol rinse, the fluorine content dropped dramatically to approximately 6%, and no phosphorus could be resolved. Results obtained for BF_4 tests show similar results. Fluorine was present in amounts near 10% and boron near 5%. After the methanol rinse fluorine dropped below 5% and boron could not be resolved. Elements including: oxygen, carbon, nitrogen, and silicon were studied as well as boron, phosphorus, and fluorine. Prior to rinsing, nitrogen and silicon were low and carbon and oxygen were high. Subsequent to rinsing, nitrogen and silicon content rose and the oxygen and carbon content declined, as the Si_3N_4 substrate became the dominant component.

XPS analysis also gives chemical state information of the sample surface. High resolution scans of the fluorine 1s region shows 2 distinct peaks for BF_4 tests. The two peaks are located at 685eV and 689eV and can be attributed to elemental fluorine and the presence of either BF_3 or BF_4, respectively. After rinsing with methanol the BF_3/BF_4 peak reduces to trace amounts. High-resolution scans of the B region strengthen the case for BF_x compounds. B1s region scans shows the presence of BF_x compounds with a peak at approximately 194 eV. The presence of B_2O_3 may also be attributed to this peak and cannot be ruled out. After rinsing with methanol only small amounts of the BF_x can be detected. Infrared spectroscopy was also employed determine chemical state information on the sample surfaces. B-F bond stretching in BF_3 was determined to be present due to a distinct peak near 1450 cm^{-1} (21). A high resolution spectrum of the F1s region for PF_6 tests shows 2 distinct peaks similar to the BF_4. The two peaks near 685 eV and 689 eV are representative of elemental fluorine and PF_x compounds, respectively. High-resolution scans of the P region strengthen the case for PF_x compounds due to the presence of a peak at 137.5 eV, which corresponds to PF_x compounds. FTIR data from the PF_6 tests were featureless. XPS scans of the F 1s region can be seen in Figure 1 (Data are from reference 7).

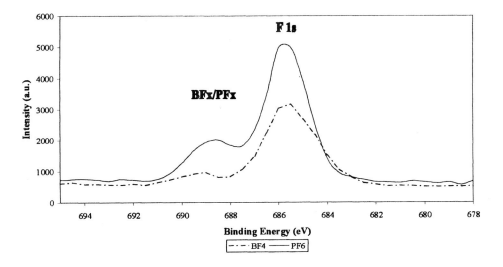

Figure 1: XPS Scan of the Fluorine 1s Region Subsequent to Tribology Testing

Tribology

Several different experiments were undertaken to determine the effects ionic liquids have as a boundary lubricant additive to water. First, water alone was run on silicon nitride to validate our experimental data with previous reports. Generally, silicon nitride in water consists of a running-in period where a tribochemical reaction takes place along with surface smoothing to achieve hydrodynamic lubrication. The results from the water only tests matched other reports of water lubricated sliding on silicon nitride reasonably well with respect to friction coefficients as well as running-in periods.

Two ionic liquid solutions were run on silicon nitride to determine their properties as a boundary lubricant additive. First, a PF_6 solution was tested with the results being very exciting. The running-in period for silicon nitride in water was greatly reduced as the friction coefficient dropped to low levels after only a few hundred cycles instead of the typical few thousand cycles. The BF_4 solution exhibited similar results as the friction coefficient dropped dramatically compared to the water alone test. Comparing the two ionic liquid solutions, it can be seen that the PF_6 solution transitions to low frictions faster than its BF_4 counterpart. The results of the tribology tests are shown in Figure 2 (Data are from reference 7).

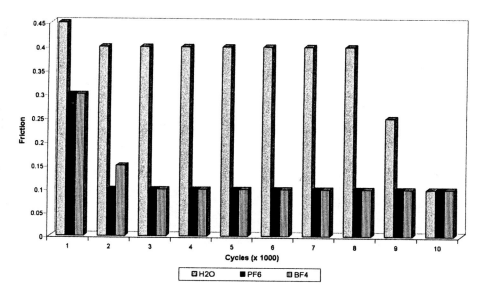

Figure 2: Ionic Liquids Effects on Si_3N_4

Discussion

The pin on disk tests on ceramics with ionic liquid additives yield very intriguing results. The friction coefficients dropped much faster during the running-in period when compared to water alone. The running-in period is reduced to only a few hundred of cycles compared to several thousand for water alone. These findings are similar to previous tribological results. Ye et al have reported low friction using and oscillating friction and wear machine. Friction values were obtained by using two drops of pure 1-methyl-3-hexylimidazoluim tetrafluoroborate which resulted in friction coefficients below 0.1 for many frictional pairs and as low as 0.065 for Si_3N_4/ sialon. Ye et al attribute the tribological properties to the decomposition of the ionic liquid into fluoride, boron oxide, and boron nitride (*4*). Other work done by Gao et al with chemically similar materials, imidazoline borates, propose the mechanism for lubrication is the reaction of the borates to form a boundary film containing nitrogen and boron compounds (6). AFRL XPS testing, for BF_4 tests, reveal that elemental B and BF_x compounds on the surface. Corresponding FTIR data shows B-F bond stretching in BF_3 compounds (21). Cumulatively, these works suggest that the mechanism by which ionic liquids perform is that the BF_4 ionic liquid breaks down to form a BF_3 film, which provides an anti-friction surface. Similarly PF_x compounds form when using PF_6 ionic liquid. On the contrary, these surface films produced by the ionic liquids do not behave similar to other anti-wear additives as both films are easily removed from the surface by rinsing with methanol. For example, Zinc dialkyldithiophosphate (ZDDP), forms a surface coating that remains after washing (22). This suggests that additional mechanisms could be controlling friction and wear properties.

A second proposed mechanism is surface etching by HF. This could enhance the surface smoothing process and therefore reduce the running-in period. To prove or disprove this mechanism a solution of 0.5% HF was tested in water with a silicon nitride/silicon nitride contact. The sample surface was monitored before and after tested and it was determined through roughness measurements that the surface did not significantly change due to the HF. Thus, surface etching by HF is not controlling the friction and wear properties.

A third proposed mechanism is the creation of an electric double layer consisting of two equal and oppositely charged regions (23,24). The electric double layer has been studied recently in a water environment (25,26). Wong et al studied the effect on a thin water lubricating film and concluded that the electric double layer played a significant role for films less than 100nm, which is larger than the water film thickness between Si_3N_4 versus itself during hydrodynamic lubrication (70nm) (8). The effects of the electric double layer could be increased the load-carrying capacity and lower friction coefficients (25). Due to the increased concentration of ionic liquids in the double layer, the fluid near the sample surface may act similar to a confined fluid where the effective viscosity increases near the surface (27). We propose that the electric double layer provides boundary layer protection as well as increased viscosity near the surface that would enhance hydrodynamic lubrication.

Summary

Ionic liquids have been studied intensively as environmentally friendly solvents. This report focuses on the potential use of ionic liquids in the field of tribology, particularly to the use of ionic liquids as a boundary lubricant additive to water. It has been shown that during water lubricated sliding, silicon nitride ceramics undergo a characteristic running-in period of high friction before transitioning to hydrodynamic lubrication. The running-in period is significantly decreased due to the addition of an ionic liquid boundary lubricant. We evaluated three mechanisms to explain this behavior. The first mechanism attributed the behavior to chemical etching of the surface by HF. This was disproved by running a dilute HF on silicon nitride, as this solution did not reduce the running-in period similar to ionic liquids. The second mechanism is the formation of a transfer layer on the surface. XPS analysis of the samples shows a presence of BF_x and PF_x compounds in the wear track area, suggesting this mechanism may be contributing to the observed behavior. The third proposed mechanism, due to the nature of the ionic liquid, is the creation of an electric double layer, which increases the load carrying capability and lowers the friction. Overall, the additions of ionic liquids to water greatly reduce the running in period of silicon nitride ceramics and consequently reduce the wear of the ceramics. Due to ionic liquids many unique properties they could be used in several other applications, such as additives to turbine engine oils (excellent thermal stability) or for space applications (negligible volatility). Ionic liquids show great potential as additives for a wide range of tribological applications.

References

1. Welton, T.; *Chem. Rev.*; **1999**, 99, 2071.
2. Holbrey, J.D; Seddon K.R.; *Clean Prod. and Processes* **1999**,1,223.
3. Hagiwara, R.; Ito, Y.; *J. Fluorine Chem.* **2000**, 105, 221.
4. Ye, C.; Liu, W.; Chen, Y.; Yu, L.; *Chem. Comm.* **2001**, 2244.
5. Ye, C.; Liu, W.; Chen, Y.; Ou,Z.; *Wear* **2002**, 253, 579.
6. Gao, Y.; Jing, Y.; Zhang, Z.; Chen, G.; Xue,Q.; *Wear* **2002**, 253, 576.
7. Phillips, B.S.; Zabinski, J.S.; Accepted to *Tribology Letters* **2004**.
8. Tomizawa, H.; Fischer, T.E.; *ASLE Transactions* **1986**, 30, 41.
9. Chen, M.; Kato, K.; Adachi, K.; *Tribology Letters* **2001**, 11, 23.
10. Wong, W.; Umehara, N.; Kato, K.; *Tribology Letters* **1998**, 5, 303.
11. Xu, J.; Kato, K.; Hirayama,T.; *Wear* **1997**, 205, 55.
12. Mizuhara, K.; *Eurotrib.* **1993**, 3, 52.
13. Lancaster, J.K.; Mashal Y.A.; A.G. Atkins, *J. Phys. D.* **1992**, 25, A205.
14. Gao, Y.; Fang, L.; Su, J.; Xie, Z.; *Wear* **1997**, 206, 87.
15. Saito, T.; Imada, Y.; Honda, F.; *Wear* **1997**, 205,153.
16. Andersson, P.; *Wear* **1992**, 154, 37.
17. Kitaoka, S.; Tsuji, T.; Yamaguchi, Y.; Kashiwagi, K.; *Wear* **1997**, 205, 40.
18. Saito, T.; Hosoe, T.; Honda, F.; *Wear* **2001**, 247, 223.
19. Loffelbein, B.; Woydt, M.; Habig, K.H.; *Wear* **1993**, 162-164, 220.
20. Cavdar, B.; Sharma, S.; Gschwender, L.; *Lubrication Engineering* **1994**, 50, 895.
21. Colthup, N.B.; Daly, L.H.; Wiberley, S.E.; *Introduction to Infared and Raman Spectroscopy,* Academic Press:New York, NY. 1964.
22. Wan, Y.; Cao, L.; Xue; Q.; *Tribology Int.* **1998**, 30, 767.

23. Hiemenz,P.C.; *Principles of Colloid and Surface Chemistry:* Marcel Dekker, New York, NY, 1986.
24. Israelachvili, J.; *Intermolecular and Surface Forces*: Academic Press, San Diego, CA, 1992.
25. Wong, P.L.; Huang, P.; Meng, Y.; *Tribology Letters* **2003**, 14, 197.
26. Xu, J.; Kato, K.; *Wear* **2000**, 245, 61.
27. Luengo, G.; Israelachvili, J.; Granick, S.; *Wear* **1996**, 200, 328.

Phase Equilibria

Chapter 20

Phase Equilibria (SLE, LLE) of *N,N*-Dialkylimidazolium Hexafluorophosphate or Chloride

U. Domańska, A. Marciniak, and R. Bogel-Lukasik

Faculty of Chemistry, Warsaw University of Technology, 00–664
Warsaw, Poland

The solubility (SLE and LLE) of 1-ethyl-3-methylimidazolium hexafluorophosphate [emim][PF$_6$] in heptane and cyclohexane and solubility (SLE) of 1-dodecyl-3-methylimidazolium chloride [C$_{12}$mim]Cl in toluene and *m*-xylene have been measured. The SLE data were correlated by means of the Wilson, UNIQUAC and NRTL equations. The review of the previous work is presented.

The solubility of 1-ethyl-3-methylimidazolium hexafluorophosphate, [emim][PF$_6$] in aromatic hydrocarbons (benzene, toluene, ethylbenzene, *o*-xylene, *m*-xylene, *p*-xylene) (*1*) and in alcohols (methanol, ethanol, propan-1-ol, propan-2-ol, butan-1-ol, butan-2-ol, *tert*-butanol and 3-methylbutan-1-ol) (*2*) as well as of 1-butyl-3-methyl-imidazolium hexafluorophosphate, [bmim][PF$_6$] in the same aromatic hydrocarbons, in *n*-alkanes (pentane, hexane, heptane, octane) and in cyclohydrocarbons (cyclopentane, cyclohexane) has been measured by a dynamic method from 290 K to the melting point of ionic liquid or to the boiling point of the solvent (*1*). The solubility of [emim][PF$_6$] in alcohols (ethanol, propan-1-ol, butan-1-ol) was presented earlier (3). The solubility of [emim][PF$_6$] and [bmim][PF$_6$] in aromatic hydrocarbons and in alcohols decreases with an increase of the molecular weight of the solvent. The differences on the solubility in o-, m- p-xylene are not significant. The solubility of [emim][PF$_6$] in alcohols, with the exception of methanol, show the mutual liquid-liquid equilibrium. Solubility is better in secondary alcohols than in primary alcohols. The shape of the equilibrium curve is similar for [emim][PF$_6$] or [bmim][PF$_6$] in every alcohol. The observations of upper critical solution temperatures were limited by the boiling temperature of the solvent. For example for [emim][PF$_6$] solubilities in methanol and ethanol are higher than that in aromatic hydrocarbons. The miscibility gap in C$_3$ alcohols is bigger than that in benzene, but comparable with the solubility in toluene; solubility in 3-methyl-butan-1-ol is very low and similar to ethylbenzene and o-, m- and p-xylene (1, 2).

Also the solid-liquid equilibria of 1-butyl- or decyl- or dodecyl-3-methylimidazolium chloride [C$_4$ or C$_{10}$ or C$_{12}$ mim]Cl in alcohols has been measured by the same method (4, 5, 6). The experimental SLE phase diagrams investigated for [C$_4$ or C$_{10}$ or C$_{12}$ mim]Cl have shown that the solubility of IL in alcohols decreases with an increase of the carbon chain of an alcohol from C$_2$ to C$_8$ with exception of [C$_4$mim]Cl in butan-1-ol. The [C$_4$mim]Cl exhibits the best solubility in 1-butanol, what can be explained by the best packing effect in the solution for the same number of carbon atoms in the solvent and butyl substituent at the imidazolium ring. Solubility of [C$_4$mim]Cl in longer chain alcohols is similar to that in octan-1-ol. The liquidus curves of the primary, secondary and tertiary alcohols exhibit similar shapes. The solubility increases in order butan-1-ol > butan-2-ol > tert-butanol (4). Positive and negative deviations from the ideality were found, thus the solubility was higher or partly higher than the ideal one in ethanol, butan-1-ol, butan-2-ol and tert-butanol and was lower than the ideal solubility in other alcohols. The complete phase diagrams [C$_4$ or C$_{10}$ or C$_{12}$ mim]Cl were found to show eutectic behavior for {[C$_4$ or C$_{10}$ or C$_{12}$ mim]Cl + tert-butanol, or decan-1-ol, or dodecan-1-ol}(4, 5, 6).

The melting and solid-solid phase transition temperatures and enthalpies were determined by the differential scanning calorimetry (DSC) for every IL (1,2,7). It is possible to conclude from the DSC results for [C$_4$ or C$_{10}$ or C$_{12}$ mim]Cl that the 1-decyl-3-methylimidazolium chloride, [C$_{10}$mim]Cl, exhibits similar behavior to 1-dodecyl-3-methylimidazolium chloride, [C$_{12}$mim]Cl. Both [C$_{10}$mim]Cl and [C$_{12}$mim]Cl exhibit solid-solid phase transitions with the difference in the enthalpy of melting. For [C$_{12}$mim]Cl the mesomorphic plastic phase with the very small heat of melting transition (0.604 kJ·mol^{-1}) exists (6). Only the next transition from plastic phase to crystalline phase is accompanied by the high heat effect equals to 23.580 kJ·mol^{-1}. On the contrary, for example the 1-butyl-3-methylimidazolium chloride, [C$_4$mim]Cl, do not exhibit any solid-solid phase transitions but glass transition at about 197 K (4).

The results of the solid-liquid correlations by the well known G^E models Wilson (8), UNIQUAC ASM (9) and NRTL1 (10) equations depend on the shape of the liquidus curves, it means on the enthalpies of solid-solid phase transitions and the equation used. The worst results were obtained for [C$_{12}$mim]Cl (6) with high solid-solid phase transition enthalpy.

From the solubilities of 1-alkyl-3-methylimidazolium chloride, [C$_n$-mim]Cl, where n = 4, 8, 10, 12 in octan-1-ol and water the octan-1-ol/water partition coefficients as a function of temperature and alkyl substituent have been measured (7). The solubility of [C$_n$-mim]Cl, where n = 10, 12 in octan-1-ol is comparable to that of [C$_4$mim]Cl in octan-1-ol. Liquid 1-octyl-3-methylimidazolium chloride, [C$_8$mim]Cl, is not miscible with octan-1-ol and water, thus consequently the liquid-liquid equilibrium, LLE was measured in this system. The differences of solubilities in water for n = 4 and n = 12 are not very significant. Additionally, the immiscibility region was observed for the higher concentration of [C$_{10}$mim]Cl in water. The intermolecular solute–solvent interaction of 1-butyl-3-methylimidazolium chloride with water is higher than for other 1-alkyl-3-methylimidazolium chlorides. Experimental partition coefficients (log P) are negative in three temperatures, what is the evidence of the possibility of using these ionic liquids as a green solvents (7).

The solubilities of 1-dodecyl-3-methylimidazolium chloride [C$_{12}$mim]Cl in hydrocarbons (benzene, octane, decane, dodecane), ethers {dipropyl ether, methyl 1,1-dimethylethyl ether (MTBE), methyl 1,1-dimethylpropyl ether (MTAE), tetrahydrofuran, (THF)} have been also measured (11). The influence of the anion {[Cl], [PF$_6$], [CH$_3$SO$_4$]} on the solubility of 1-butyl-3-methylimidazolium salts in dipropyl ether was tested as well (11).

The solubilities, liquid-liquid equilibrium data for ternary mixtures of {[C$_8$mim]Cl + benzene + an alkane (heptane, dodecane, hexadecane)} at 298.15 K were presented nowadays (12). The effectiveness of the extraction of benzene from an alkane by IL was reported as a ratio of solubilities in the two phases and as the "selectivity". The last one was found to increase with increasing carbon number of the alkane (12).

Ionic liquids are not only important in Green Chemistry because they create a cleaner and more sustainable chemistry but they also show huge advantage in extraction processes comparing with traditional substances as sulfolane or NMP (13, 14). It was shown that the large miscibility gap exists in the mixtures of 1-ethyl-3-methylimidazolium bis(trifluoromethylsulfonyl) imide, [emim][$CF_3SO_2)_2N$] or 1-ethyl-3-methylimidazolium ethylsulfate, [emim][$C_2H_5OSO_3$] with aliphatic (cyclohexane) and aromatic (benzene) hydrocarbons. The excess molar enthalpy for the benzene + [emim] [$CF_3SO_2)_2N$] was highly negative (about −750 J·mol^{-1}) and for the cyclohexane + [emim][$CF_3SO_2)_2N$] was positive (about 450 J·mol^{-1}). The large miscibility gap at low and high temperatures (up to the boiling point of the solvent) can directly be used for the separation of aromatic from aliphatic hydrocarbons by liquid-liquid extraction (13). ILs, [C_2 or C_4 mim][BF_4] have been shown to exhibit many properties which allow for their application as selective solvents in solvent extraction. The addition of these ILs result in the braking of the azeotropic system behaviour of (THF + water) or (ethanol + water) (14). Therefore, IL represent a promising class of highly selective entrainers, whose properties can be tailored to the individual application.

The liquid-liquid equilibria, the influence of high pressure up to 800 bar on the IL/water system, or IL/ deuterated water system and the solvent effect in the phase diagram of ternary mixture {[C_4mim][PF_6] + (ethanol + water)} was presented (15, 16).

The purpose of this paper is to report the solubility of 1-ethyl-3-methylimidazolium hexafluorophosphate, [emim][PF_6] in cyclohexane and hexane and of 1-dodecyl-3-methylimidazolium chloride [C_{12}mim]Cl in hydrocarbons (toluene and *m*-xylene) and to compare these data with published earlier hydrocarbons, alcohols and ethers (1-2, 4-7, 11). The data presented here will be useful for testing new possibilities of using ILs in extraction processes.

Experimental

The ionic liquids [emim][PF_6] and [C_{12}mim]Cl, produced by Solvent Innovation GmbH, Köln, Germany were used as received. The purity was ≥ 98 mass percent for every IL. All solvents were produced by Sigma-Aldrich Chemie GmbH, Steinheim, Germany. The solvents were fractionally distilled over different drying reagents to a mass fraction purity better than 99.8 mass percent. Liquids were stored over freshly activated molecular sieves of type 4A (Union Carbide).

Solid solubilities have been determined using a dynamic (synthetic) method from 270 K to the melting point of salt or to the boiling point of the solvent, described in detail previously (1-2, 4-7). Mixtures of solute and solvent were prepared by weighing the pure components to within $1 \cdot 10^{-4}$ g. The sample of solute and solvent were heated very slowly (at less than 2 K·h^{-1} near the equilibrium temperature) with continuous stirring inside a Pyrex glass cell, placed in a thermostat. The crystal disappearance temperatures, detected visually, were measured with a calibrated GALLENKAMP AUTOTHERM II thermometer totally immersed in the thermostating liquid. Measurements were carried out over a wide range of solute mole fraction from 0 to 1. The accuracy of temperature measurements was ± 0.01 K while the error in the mole fraction did not exceed δx_1 = 0.0005. Every experimental point was obtained from a new sample. Physical properties and DSC data of pure substances were discuss earlier (1-2, 4-7).

Results and Discussion

The solubilities of [emim][PF$_6$] in two different hydrocarbons are shown in Tables 1, 2 and these of [C$_{12}$mim]Cl in two aromatic hydrocarbons in Tables 3, 4. Tables 3 and 4 include the direct experimental results of the SLE temperatures, T_{1SLE} (for two crystallographic forms α, β and γ) and T_{1LLE} versus x_1, the mole fraction of the IL in the saturated solution for the investigated systems. The liquid–liquid equilibria, LLE were observed for [emim][PF$_6$] in cyclohexane and heptane as before in aromatic hydrocarbons (1) and in alcohols (2). The intermolecular solute-solvent interactions are not significant when the binary liquid phases are observed. It is not surprising for hydrocarbons. The differences in solubilities in heptane for [emim][PF$_6$] and [bmim][PF$_6$] (1) are small. Basic [emim][PF$_6$] can act both as hydrogen-bond acceptor and donor and would be expected to interact with polar solvents, such as alcohols. However, the mutual solubility of [emim][PF$_6$] and [bmim][PF$_6$] in alcohols was observed. The liquidus curves exhibit different shapes for hydrocarbons and alcohols with the lower area of the binary liquids for alcohols. See Figure 1 for cyclohexane and propan-1-ol. The mutual liquid-liquid solubility of [emim][PF$_6$] increases in the order: benzene < propan-1-ol < dipropyl ether < heptane (see Figure 2). The [bmim][PF$_6$] is more soluble in aromatic compounds than [emim][PF$_6$] and the mutual solubility in benzene and its alkyl derivatives decreases with an increase of the alkyl subtituent at the benzene ring (*1*). Opposite influence of the substituent is observed for [C$_{12}$mim]Cl. This IL is more soluble in toluene and *m*-xylene than in benzene {this work and (*11*)}. The observations of the upper critical solution temperatures (USCT) were limited by the boiling temperature of the solvent. In this work and published before (1, 2) it was impossible to detect by the visual method the mutual solubility of hexafluorophosphate salts in the solvent rich phase. The spectroscopic or other technique is necessary to use. Comparison of the solubilities of three ILs in aromatic compounds, toluene and *m*-xylene is shown in Figures 3 and 4. The binary mixtures of the [C$_{12}$mim]Cl with aromatic compounds may be assumed as a simple eutectic systems.

Table 1. Experimental Solid-Liquid Equilibrium Temperatures T_{1SLE} and Liquid-Liquid Temperatures T_{1LLE} for {x_1 [emim][PF$_6$] + (1- x_1) Cyclohexane} System

x_1	T_{1SLE}^a	x_1	T_{1SLE}^a	T_{1LLE}^a
0.0025	331.85	0.6854	331.85	
0.0045	331.85	0.7347	331.85	
0.0194	331.85	0.7715	331.85	
0.0400	331.85	0.7945	331.85	
0.0847	331.85	0.8112	331.85	
0.1238	331.85	0.8240	331.85	
0.1753	331.85	0.8292	331.85	
0.2130	331.85	0.8485	331.85	352.85
0.2597	331.85	0.8617	331.85	343.29
0.3091	331.85	0.8701	331.85	337.45
0.3486	331.85	0.8735	331.85	334.97
0.3904	331.85	0.8974	331.99	
0.4315	331.85	0.9185	332.22	
0.4702	331.85	0.9282	332.29	
0.5105	331.85	0.9459	332.54	
0.5549	331.85	0.9551	332.64	
0.6016	331.85	0.9711	332.75	
0.6387	331.85	1.0000	332.80	

^{a} Units of temperature (K).

Table 2. Experimental Solid-Liquid Equilibrium Temperatures T_{1SLE} and Liquid-Liquid Temperatures T_{1LLE} for $\{x_1$ [emim][PF$_6$] + (1- x_1) Heptane$\}$ System

x_1	$T_{1SLE}{}^a$	x_1	$T_{1SLE}{}^a$	$T_{1LLE}{}^a$
0.0052	332.33	0.6159	332.33	
0.0195	332.33	0.6639	332.33	
0.0269	332.33	0.7154	332.33	
0.0456	332.33	0.7784	332.33	
0.0768	332.33	0.8316	332.33	
0.1124	332.33	0.8756	332.33	
0.1547	332.33	0.9025	332.33	
0.2070	332.33	0.9069	332.33	370.95
0.2512	332.33	0.9259	332.33	362.09
0.3026	332.33	0.9430	332.33	349.20
0.3568	332.33	0.9513	332.33	338.49
0.4026	332.33	0.9577	332.33	335.05
0.4389	332.33	0.9706	332.33	
0.4826	332.33	0.9832	332.42	
0.5242	332.33	1.0000	332.80	
0.5687	332.33			

^{a} Units of temperature (K).

Table 3. Experimental Solid-Liquid Equilibrium Temperatures T_{1SLE} for $\{x_1$ [C$_{12}$mim]Cl (for α, β and γ Crystallographic Forms) + (1- x_1) Toluene$\}$ System

x_1	$T_{1\gamma}{}^a$	$T_{1\beta}{}^a$	x_1	$T_{1\alpha}{}^a$
0.3565	273.77		0.8649	320.52
0.4042		287.86	0.8984	322.79
0.4303		294.33	0.9561	337.13
0.4567		299.09	0.9800	363.56
0.5088		305.63	1.0000	369.78
0.5642		313.15		
0.5886		314.51		
0.6347		315.08		
0.6744		315.25		
0.7582		315.92		
0.8198		318.65		

^{a} Units of temperature (K).

Table 4. Experimental Solid-Liquid Equilibrium Temperatures T_{1SLE} for $\{x_1$ [C$_{12}$mim]Cl (for α, β and γ Crystallographic Forms) + (1- x_1) m-Xylene} System

x_1	$T_{1\gamma}{}^a$	$T_{1\beta}{}^a$	x_1	$T_{1\alpha}{}^a$
0.3688	277.06		0.8560	322.19
0.4108		286.91	0.9216	330.28
0.4512		294.81	1.0000	369.78
0.5172		301.26		
0.5923		308.24		
0.6341		311.17		
0.6822		313.80		
0.7224		317.15		
0.7647		319.07		
0.8067		319.12		

^{a} Units of temperature (K).

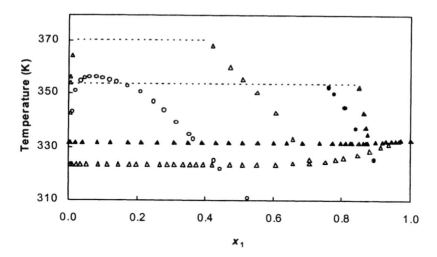

Figure 1. Solid-liquid and liquid-liquid equilibrium diagrams for $\{x_1$ an IL + $(1-x_1)$ propan-1-ol}: 0, [bmim][PF$_6$] (3) and ?, [emim][PF$_6$] (2), or $\{x_1$ an IL + $(1-x_1)$ cyclohexane}: ?, [bmim][PF$_6$] (1) and $\wedge$, [emim][PF$_6$]; dotted line, boiling temperature of a solvent.

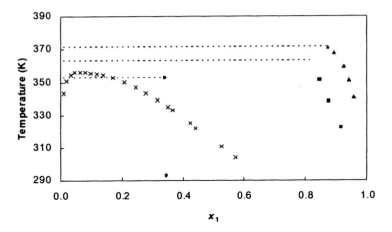

Figure 2. Liquid-liquid equilibrium diagrams for {x₁[bmim][PF₆] + (1-x₁) a solvent}: x, propan-1-ol (3), or ?, benzene (2), or ¦, dipropyl ether (11), or ^, heptane (1); dotted line, boiling temperature of a solvent.

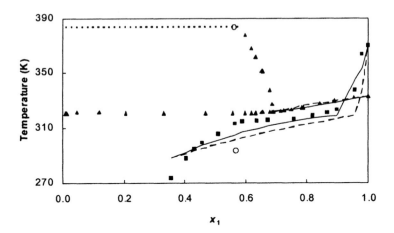

Figure 3. Solid-liquid and liquid-liquid equilibrium diagrams for {x₁ an IL + (1-x₁) toluene}: ^, [emim][PF₆] (1), or 0, [bmim][PF₆] (1), or ¦, [C₁₂mim]Cl; solid lines, calculated by the WILSON equation; dashed line, ideal solubility; dotted line, boiling temperature of a solvent.

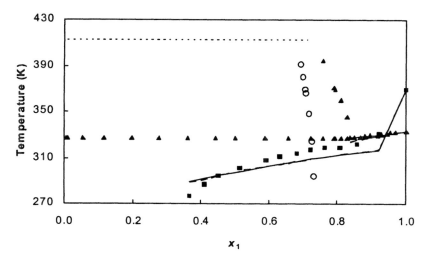

Figure 4. Solid-liquid and liquid-liquid equilibrium diagrams for {x₁ an IL + (1-x₁) m-xylene}: ^, [emim][PF₆] (1), or 0, [bmim][PF₆] (1), or ¦, [C₁₂mim]Cl; solid line, calculated by the NRTL equation; dashed line, ideal solubility; dotted line, boiling temperature of a solvent.

Solid-Liquid Equilibria Correlation

The solubility of a solid 1 showing solid-solid phase transitions in a solid phase may be express in a very general way by eq 1

$$-\ln x_1 = \frac{\Delta_{fus}H_1}{R}\left(\frac{1}{T_1}-\frac{1}{T_{fus,1}}\right)+\frac{\Delta_{trI}H_1}{R}\left(\frac{1}{T_1}-\frac{1}{T_{trI,1}}\right)+\frac{\Delta_{trII}H_1}{R}\left(\frac{1}{T_1}-\frac{1}{T_{trII,1}}\right)-\frac{\Delta_{fus}Cp_1}{R}\left(\ln\frac{T_1}{T_{fus,1}}+\frac{T_{fus,1}}{T_1}-1\right)+\ln\gamma_1 \quad (1)$$

where x_1, γ_1, $\Delta_{fus}H_1$, $\Delta_{fus}Cp_1$, $T_{fus,1}$ and T_1 are the mole fraction, activity coefficient, enthalpy of fusion, difference in solute heat capacity between the solid and liquid at the melting point, melting temperature of the solute (1) and equilibrium temperature, respectively. $\Delta_{trI}H_1$, $\Delta_{trII}H_1$, and $T_{trI,1}$, $T_{trII,1}$ are the enthalpies of solid-solid phase transition and transition temperatures of the solute, respectively. The solubility equation for temperatures below that of the phase transition must include the effect of the transition. Equation (1) may be used assuming simple eutectic behavior with full miscibility in the liquid phase and immiscibility in the solid phase, as discussed previously (1-2). In this work the big difference between $T_{trl,1}$(DSC exp) = 310.15 K and $T_{trl,1}$(SLE exp) = 318.90 K for [C$_{12}$mim]Cl was noted, which is difficult to interpret from the thermodynamic point of view. The solid-solid phase transition temperature is the pure solid property and suppose to be not influenced by the solvent. The transition in n-alkanes are usually related to rotation of molecular chains about their long axes (here C$_{12}$). The possible explanation of this problem for [C$_{12}$mim]Cl was proposed in our previous work (6). Because of the lack of appropriate data representing the difference $\Delta_{fus}Cp_1$ between the heat capacities of the solute in the solid and the liquid states for the calculated mixtures of [C$_{12}$mim]Cl, the simplified version of the solubility eq 1 without the $\Delta_{fus}Cp_1$ term was applied. Melting temperatures were 332.80 K and 369.78 K; enthalpies of melting were 17.860 kJ·mol^{-1} and 369.79 kJ·mol^{-1} for [emim][PF$_6$] (1) and [C$_{12}$mim]Cl (6), respectively. Enthalpies of solid-solid phase transitions for [C$_{12}$mim]Cl were published earlier (6) and are equal to $\Delta_{trI}H_1$ = 23.580 kJ·mol^{-1}, $\Delta_{trII}H_1$ = 1.157 kJ·mol^{-1}. The second solid-solid phase transition temperature $T_{trII,1}$ for [C$_{12}$mim]Cl was 283.21 K (from DSC) (6). The enthalpy of fusion for [emim][PF$_6$] was $\Delta_{fus}H_1$ = 17.862 kJ·mol^{-1} (1).

In this study three methods describing the Gibbs excess energy of mixing, (G^E) are used to correlate the experimental data: the Wilson (8), UNIQUAC (17) and NRTL (18) equations. Table 5 shows the parameters and deviations for the equations used. Also the activity coefficients of the IL in binary mixture, calculated by the difference of the eq 1 and the ideal solubility equation (without the lnγ_1 term) can be discuss. The activity coefficient of the [C$_{12}$mim]Cl in the toluene and m-xylene saturated solutions are partly lower than one (solubility higher than ideal solubility for the low concentration of solute) and partly higher than one. The maximum value of activity coefficient is γ_1 = 1.45. For [emim][PF$_6$] the activity coefficient is close to one along to calculated curve.

Two adjustable parameters of the equations were found by an optimization technique using Marquardt's or Rosenbrock's maximum likelihood method of minimization:

$$\Omega = \sum_{i=1}^{n} \left[T_i^{exp} - T_i^{cal}(x_{1i}, P_1, P_2) \right]^2 \tag{2}$$

where Ω is the objective function, n is the number of experimental points and T_i^{exp} and T_i^{cal} denotes respectively experimental and calculated equilibrium temperature corresponding to the concentration x_{1i}, P_1 and P_2 are model parameters resulting from the minimization procedure. The root-mean-square deviation of temperature was defined as follows:

$$\sigma_T = \left(\sum_{i=1}^{n} \frac{\left(T_i^{exp} - T_i^{cal}\right)^2}{n-2} \right)^{\frac{1}{2}} \tag{3}$$

Parameters r_i and q_i of the UNIQUAC and NRTL models were calculated with following relationships:

$$r_i = 0.029281 V_m \tag{4}$$

$$q_i = \frac{(Z-2)r_i}{Z} + \frac{2(1-l_i)}{Z} \tag{5}$$

where V_m is the molar volume of pure component i at 298.15 K. The molar volume of solute V_m (298.15) was calculated by the group contribution method (19) and was assumed to be 242.6 cm^3·mol^{-1} and 325.8 cm^3·mol^{-1} for [emim][PF$_6$] and [C$_{12}$mim]Cl, respectively. The coordination number Z was assumed to be equal to 10 and the bulk factor l_i was assumed to be equal to 1.

The correlation for three used equations was presented with the developing of two adjustable parameters. In this work, the value of parameter α, a constant of proportionality similar to the non-randomness constant of the NRTL equation was $\alpha = 0.3$. Figures 3 and 4 show the solid-liquid phase diagram for the {[C$_{12}$mim]Cl + toluene or m-xylene} correlated by the Wilson and NRTL equations, respectively, as an examples. For [emim][PF$_6$] the results in cyclohexane are presented in Table 5 with the average standard mean deviation, $\sigma_T = 0.15$ K. For two systems presented for [C$_{12}$mim]Cl the description of solid–liquid equilibrium was given by the standard mean deviation, $\sigma_T = 6\text{-}8$ K. None of these models correlate the data successfully, inherently because the solute is a very complicated and highly interacting molecule. However, the main problem for the equation's used is the untypical shape of the [C$_{12}$mim]Cl liquidus curve and high enthalpy of the first solid-solid phase transition (much higher than the enthalpy of melting). The correlation of the LLE is at the moment impossible because of the lack of the left equilibrium line.

Planning new synthesis or liquid/liquid or solid extraction process by changing IL or only cations or organic solvent in binary mixture, many different new possibilities can be obtained from our results.

Table 5. Correlation of the Solubility Data (SLE) of {[C$_{12}$mim]Cl or [emim][PF$_6$] (1) + Hydrocarbon (2)} Mixtures, by means of the Wilson, UNIQUAC and NRTL Equations: Values of Parameters and Measures of Deviations

System	Parameters[a]		
	Wilson	UNIQUAC	NRTL[b]
	g_{21}-g_{11}/ g_{12}-g_{22}	$\Delta u_{21}/\Delta u_{12}$	$\Delta g_{21}/\Delta g_{12}$
[emim][PF$_6$]	-3829.59	11325.49	18982
+ C$_6$H$_{12}$	24142.34	-4251.60	-7278.51
[C$_{12}$mim]Cl +	-5303.08	8312.24	12836
Toluene	15621.81	-2997.16	-4177.34
[C$_{12}$mim]Cl	-679.24	-379.59	970.34
+ m-Xylene	1145.4	1465.84	-671.83
	Deviations[d]		
Solvent	Wilson	UNIQUAC	NRTL
	σ_T	σ_T	σ_T
C$_6$H$_{12}$	0.05	0.18	0.23
Toluene	6.64	6.72	7.16
m-xylene	7.54	8.33	7.54

[a] Units are (J·mol^{-1}).

[b] Calculated with the third non-randomness parameter $\alpha = 0.3$.

[d] Units are (K).

References

1. Domańska, U.; Marciniak, A. *J. Chem. Eng. Data* **2003**, *48*, 451-456.

2. Domańska, U.; Marciniak, A. *J. Phys. Chem. B* **2003**, *108*, 2376-2382.

3. Marsh, K. N.; Deev, A.; Wu, A. C. –T.; Tran, E.; Klamt, A. *Kor. J. Chem. Eng.* **2002**, *19*, 357-362.

4. Domańska, U.; Bogel-Łukasik, E. *Fluid Phase Equilib.* **2004**, *218*, 123-129.

5. Domańska, U.; Bogel-Łukasik, E. *Ind. Eng. Chem. Res.* **2003**, *42*, 6986-6992.

6. Domańska, U.; Bogel-Łukasik, E.; Bogel-Łukasik, R. *J. Phys. Chem. B* **2003**, *107*, 1858-1863.

7. Domańska, U.; Bogel-Łukasik, E.; Bogel-Łukasik, R. *Chem. Eur. J.* **2003**, *9*, 3033-3049.

8. Wilson, G. M. *J. Am. Chem. Soc.* **1964**, *86*, 127-130.

9. Nagata, I. *Fluid Phase Equilib.* **1985**, *19*, 153-174.

10. Nagata, I.; Nakamiya, Y.; Katoh, K.; Kayabu, J. *Thermochim. Acta* **1981**, *45*, 153-165.

11. Domańska, U.; Mazurowska, L. *Fluid Phase Equilib.* **2004** in press.

12. Letcher, T. M.; Deenadayalu, N. *J. Chem Thermodyn.* **2003**, *35*, 67-76.

13. Krummen, M.; Gmehling, J. 17th IUPAC Conference on Chemical Thermodynamics, Rostock, Germany, **2002**.

14. Jork, C.; Seiler, M.; Arlt, W. *Thermodynamics 2003*, Cambridge, UK, **2003**.

15. Najdanovic-Visac, V.; Esperanca, J. M. S. S.; Rebelo, L. P. N.; Nunes da Ponte, M.; Guedes, H. J. R.; Pires, P. F.; Szydlowski, J. 19th European Seminar on Applied Thermodynamics, Santorini, Greece, **2002**.

16. Rebelo, L. P. N.; Najdanovic-Visac, V.; Visak, Z. P.; Nunes da Ponte, M.; Tronoso, J.; Cerdeirina, C. A.; Romani, L. *Phys. Chem. Chem. Phys.* **2002**, *4*, 2251-2261.

17. Abrams, D. S.; Prausnitz, J. M. *AIChE J.* **1975**, *211*, 116-128.

18. Renon, H.; Prausnitz, J. M. *AIChE J.* **1968**, *14*, 135-144.

19. Barton, A. F. M. *CRC Handbook of Solubility Parameter*; CRC Press: Boca Raton, Fl, 1985.

Chapter 21

Phase Behavior and Thermodynamic Properties of Ionic Liquids, Ionic Liquid Mixtures, and Ionic Liquid Solutions

L. P. N. Rebelo *, V. Najdanovic-Visak, R. Gomes de Azevedo,
J. M. S. S. Esperança, M Nunes da Ponte, H. J. R. Guedes,
Z. P. Visak, H. C. de Sousa, J. Szydlowski, J. N. Canongia Lopes,
and T. C. Cordeiro

Instituto de Tecnologia Química e Biológica, ITQB 2, Universidade Nova
de Lisboa, Av. da República, Apartado 127, 2780–901 Oeiras, Portugal
*Corresponding author: luis.rebelo@itqb.unl.pt

An overview of experimental and theoretical studies recently performed in the Oeiras/Lisbon laboratories is provided. Typical showcase examples of UCST demixing of ionic liquid solutions are presented and discussed. Co-solvency, pressure, and isotope effects were investigated. In order to rationalize the observed effects, a phenomenological g^E-model was successfully applied, which permitted us to establish strong links between phase behavior and excess properties. Speed of propagation of ultrasound waves and densities in pure ILs as a function of temperature and pressure were determined from which several other thermodynamic properties such as compressibilities, expansivities and heat capacities, were derived. The quasi-ideality of mixtures of ILs, as judged by the small values of their excess volumes, could have been predicted by the master linear representation of their pure liquid volumes as the size of either the cation or the anion change. Research has been carried out at a broad range of pressures, typically up to 1600 bar, sometimes inside the metastable liquid region. The current study focuses on $[C_nmim][PF_6]$, $[C_nmim][NTf_2]$, and $[C_nmim][BF_4]$ where n is usually 4, but generally $2 < n < 10$.

Introduction

In spite of the increasing attention that room temperature ionic liquids (RTILs) has recently received in respect to their use in synthesis and catalysis, little about their physical properties and phase behavior in solution is known (1).

Within the RTILs class of compounds, those based on the cation 1-alkyl-3-methylimidazolium ($[R_n mim]^+$) are among the most popular and commonly used. Hexafluorophosphate, $[PF_6]^-$, and tetrafluorborate, $[BF_4]^-$, based RTILs are historically the most important and the most commonly investigated. We have thus chosen both for our studies. However, they can undergo hydrolisis producing HF in contact with water (2), mainly at high temperatures (3). An alternative anion has thus been considered – bis(trifluoromethylsulfonyl)amide ($[N\{SO_2(CF_3)\}_2]^-$, or $[NTf_2]^-$). RTILs also show great potential as possible extractants of a wide variety of components in aqueous solution media. In particular, it is predicted that they may come to play an important role in the recovery of butanol and ethanol produced in fermentation processes (4). In regard to liquid-liquid phase behavior, the current study focuses on solutions of [bmim][PF_6] + ethanol (EtOH) and/or water, on those of [bmim][NTf_2] + 2-methylpropanol (i-BuOH) and/or water, and on [bmim][BF_4] + water because they present partial immiscibility not far from room temperature, and total miscibility at higher temperatures (upper critical solution temperature – UCST), making them attractive from a technological perspective. In the search for a more thorough understanding of the phase behavior of RTIL solutions, we have followed the phase diagram shifts as a function of pressure and isotope effects. Phase diagrams will be discussed using an approach based on a phenomenological G^E-model (5) coupled with the statistical-mechanical theory of isotope effects (6).

Measurements of the speed of sound (SS), u, in liquids have proven to constitute a powerful source of valuable information about the thermophysical properties of chemical substances and their mixtures (7,8), especially if speed of sound data can be combined with those of density. Such a combination allows one to calculate other physical properties of ILs such as isoentropic (κ_s) and isothermal (κ_T) compressibilities, isobaric thermal expansivities (α_p), isobaric (c_p) and isochoric (c_v) specific heat capacities and thermal pressure coefficients (γ_v). To the best of our knowledge, these are the first measurements of SS of ILs.

If one can, in principle, conceive of the existence of about 10^6 pure ILs, this figure is augmented to $\sim 10^{12}$ if binary combinations are taken into account (1b). Studies of the physical behavior of binary (IL(1) + IL(2)) mixtures are thus also important. Interesting situations arise from mixing two ILs which share either a common cation or a common anion. For instance, in the first case, one observes excess properties emerging from the mutual interaction of two distinct anions in a constant cation's field. It is theoretically possible that in some selected cases the excess molar Gibbs energy could reach a critical value and phase separation would thus be triggered.

Experimental

Chemicals and Preparation of Solutions

The ILs [bmim][PF_6] and [$C_n$mim][NTf_2], where n = 2, 4, 6, 8 ,10, were synthesized and purified at the QUILL Centre, Belfast, according to procedures found elsewhere (9). They were washed several times with water to decrease the chloride content. It was checked that no precipitation (of AgCl) would occur by addition of $AgNO_3$ to the wash water. Their purity (estimated at 99.8%) was checked by NMR spectroscopy. [bmim][BF_4] (purity > 98%) was purchased from Solvent Innovations. Vacuum and moderate temperature (60 °C) were applied to all the RTIL samples for several days, in order to reduce the water content to a negligible value for present purposes. High-quality ethanol (max. 0.02 % of water content as claimed by the manufacturer) was purchased from *Panreac* and 2-methylpropanol from Riedel-de-Haen (better than 99.0% purity). Both solvents were further dried with 3 Å molecular sieves. Monodeuteriated ethanol, CH_3CH_2OD, from Aldrich (better than 99.5 % atom D) was also further dried with 3 Å molecular sieves. Water was distilled and deionized using a Milli-Q water filtration system from Millipore. Heavy water was a donation of the KFKI, Hungary, and arrived with a stated purity of 99.84 % isotopic content. All ionic liquid solutions and ionic liquid mixtures were gravimetrically prepared to an estimated uncertainty of 0.02 % for a typical non-diluted mass percentage.

Equipment

A He-Ne laser light-scattering glass capillary cell (internal volume ≈ 1.0 cm^3, optical length ≈ 2.6 mm) was used for the accurate detection of cloud-points. The apparatus, as well as the methodology used for the determination of phase transitions, have recently been described in detail (10). Cloud-point temperature accuracy is typically ±0.01 K in the range 240 < T/K < 400. As for pressure, accuracy is ±0.1 bar up to 50 bar. In the case of experiments where pressure was raised above 50 bar (and up to 700 bar), a novel sapphire/stainless steel cell (*11*) replaced the original glass capillary one. The total volume of injected solution is typically 1.6 cm^3, although the optical volume roughly corresponds to a mere 0.5 cm^3. In the case of isothermal runs, cloud-point temperature accuracy was maintained (±0.01 K) but it worsened a bit for isobaric runs. As for pressure, the uncertainty is ±1 bar in this higher-pressure range.

In order to measure the speed of propagation of sound waves in liquids using a non-intrusive method, a new cell was designed and built (Figure 1),

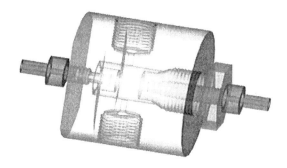

Figure 1. Non-intrusive microcell for sound-speed measurements in ILs.
The ultrasonic transducers are not in direct contact with the liquid under study.
(See page 3 of color insert.)

which is relatively similar to another that we recently developed (*12*). The total internal volume of the cell (and, the liquid) is very small (less than 0.8 cm^3). The cell is capable of reaching pressures of the order of 2000 bar. It was calibrated in the following temperature, pressure, and speed of sound ranges: $278 < T/K < 338$; $1 < p/\text{bar} < 1600$; $1000 < u/\text{ms}^{-1} < 1850$.

Densities, ρ, in the temperature range 298 K to 340 K and pressure range 1 bar to 600 bar were measured using a previously calibrated (*12*) *Anton Paar* DMA 512P vibrating tube densimeter, where temperature is controlled to ± 0.01 K and pressure accuracy and precision are better than 0.05%. Three sets of calibrating fluids corresponding to three density ranges are currently in use. The overall density precision is typically 0.002 %, while its estimated uncertainty (judging by the residuals of the overall fit in comparison with literature data for the calibrating liquids) is 0.02%.

Results and Discussion

Phase Diagrams

By and large, the information so far accumulated on L-L phase diagrams that depict phase separation of solutions of IL + solvent *(3,13)*, permits us to classify the phase diagrams as belonging to two major types (see Figure 2): (i) partially immiscible and (ii) UCST. In the first case (which in a certain manner resembles that of the hour-glass type of phase diagrams), the situation most commonly encountered is one in which the solvent dissolves to a certain extent in the IL, whereas the IL pratically does not dissolve in the solvent. Also, the solvent dissolves better in the IL as temperature rises. In the UCST-type of phase diagrams, there is partial mutual solubility at low temperatures, which transforms itself into total mutual miscibility above a certain higher temperature (critical temperature). Both types of diagrams have in common the fact that the envelope defining the two-phase region is centered at relatively low values of mole-fractions of IL.

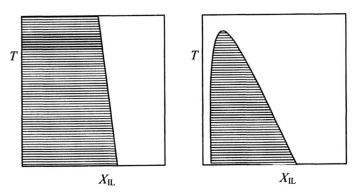

Figure 2. Two types of phase diagrams of (IL + solvent) systems. X_{IL} is the mole fraction of IL. Two-phase regions are indicated by the hatched areas.

It is possible that these "two types" of phase diagrams do in fact belong to the same class, because either vaporization of the solvent and/or the degradation of the IL as temperature raises may prevent us from detecting a potentially existent critical point at high temperatures. Interestingly, to the best of our knowledge, UCST-type phase diagrams in IL + solvent so far have only been found for cases where the solvent is either water or an alcohol.

Table I reports the coordinates of the critical point and the parameters which define the two phase envelopes according to the fit to experimental cloud-point values provided by the following scaling-type equation for three model cases of UCST behavior.

$$|w - w_c| = A \cdot \left(\frac{T_c - T}{T_c} \right)^{\beta} \tag{1}$$

where w and the subscript c refer to the weight fraction of IL and critical conditions, respectively.

TABLE I: Parameters of the scaling-type Eq. (1) for a nominal pressure of 1 bar of model cases where UCST behavior is observed

	[bmim][PF$_6$] + ethanol	[bmim][NTf$_2$] + 2-methylpropanol	[bmim][BF$_4$] + water
A	0.789	1.010	0.580
β	0.295	0.257	0.218
T_c/K	324.8	303.5	277.6
w_c	0.42	0.43	0.49
x_c [a]	0.11	0.12	0.07

[a] x_c is the mole fraction of IL at the L-L critical point.

Co-solvent, Pressure, and Isotope Effects

One of the most striking features that emerged from these studies was the observation of a very strong co-solvent effect between water and alcohols when they are mixed with an IL. The magnitude of a co-solvent effect can be defined as the difference between the mean value of the critical temperatures of the two binary mixtures (IL-solvent(1) and IL-solvent(2)) and the actual value of the lowest transition temperature found in the ternary system. In the case where IL = [bmim][PF$_6$], we found (3) a co-solvent effect of 80 K for water-ethanol as the mixed solvent. For [bmim][NTf₂] in water + 2-methylpropanol the effect must certainly be even larger although it cannot be precisely quantified as the experimental UCST of [bmim][NTf₂] + water is unknown. Nonetheless, it is expected that the T_c of the latter system should be higher than that of [bmim][PF₆] + water ($T_c \sim 410$ K) because the mutual solubility of this IL with water is lower than that of [bmim][PF₆] + water (*1d*). The effect of the addition of water to [bmim][NTf₂] + 2-methylpropanol is illustrated in Fig. 3 (the minimum corresponds roughly to a situation where the ratio (mole) water/alcohol is equal to 2/3). These findings have two implications: (i) the presence of water dramatically affects the binary phase diagram of an IL + alcohol, and (ii) one can fine-tune one-phase versus two-phase conditions at constant temperature by manipulating the water/alcohol ratio.

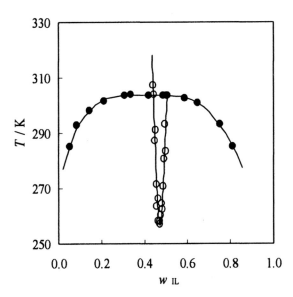

Figure 3. Phase diagram of [bmim][NTf₂] + 2-methylpropanol (•). The open circles denote the change in the transition temperature and composition as water is added to the system.

In contrast, L-L critical temperature shifts by manipulation of the applied pressure seem to be generally mild. Table II reports their magnitude and sign. As will be discussed below, the sign is intimately related to the sign of the excess volume upon mixing and the (low) magnitude is dictated by the small value of V^E in comparison with that of H^E. It is interesting to note that, while in the case of alcohols as solvents pressure helps miscibility ($V^E < 0$, dT/dp < 0), the opposite occurs for water ($V^E > 0$, dT/dp > 0).

Table II: Values of $10^3 \cdot (dT(K)/dp(bar))$ of the L-L transition temperatures of selected binary systems at a nominal pressure of 1 bar

Solvent	[bmim][PF₆]	[bmim][NTf₂]	[bmim][BF₄]
+ ethanol	- 19.9		
+ H₂O	4.96		3.1 [a]
+ ethanol-H₂O (1:1)	~ 0		
+ 2-methylpropanol		- 3.0	

(a) at 300 bar

In the case of isotopic substituion in the solvent, one again observes that alcohols behave differently from water. While deuteration in water leads to a *c.a.* 4 K upwards shift in the critical temperature of [bmim][BF₄] + water, a negative shift of about 1.1 K is observed (*13a*) upon deuteration in the –OH group of ethanol (+ [bmim][PF₆]). In the latter case, it was possible to conclude that this behavior can be rationalized by a combined effect of a red shift of -15 cm^{-1} for the O-H deformation (in the COH plane) mode of ethanol with a blue shift of +35 cm^{-1} for the O-H stretching one, both upon liquid infinite dilution in the ionic liquid. These shifts are smaller, but commensurate with those found for ethanol between pure liquid and pure gas states (*14*). This analysis was recently performed (*13a*) using the mechanical-statistical theory of isotope effects applied to binary liquid mixtures.

A Phenomenological g^E-Model and the Pressure Dependence of the Critical Temperature

Some years ago we developed a modified Flory-Huggins model (*5,11,15*) based on that developed by Qian et al. (*16*) which takes into account polydispersity effects in polymer + solvent mixtures. Due to the clear analogies found between phase diagrams of polymeric solutions and those of ionic liquid solutions (see Fig. 1), we have decided to apply the same model in the latter case. Obviously, the polydispersity index is taken as 1.0, and the IL is treated as if it were a molecular entity. The model relies on the following change of the molar Gibbs energy of the binary system upon mixing,

$$\Delta G_{\mathrm{m}}/\mathrm{Jmol}^{-1} = RT\left[x_1 \ln \varphi_1 + x_2 \ln \varphi_2 + \chi(\mathrm{T})x_2 \varphi_1\right] \qquad (2)$$

and has been described in detail elsewhere (15). x, φ, and χ represent, respectively, mole fraction, segment fraction (usually similar to volume fraction), and energy interaction parameter. The latter is given by,

$$\chi(\mathrm{T},\mathrm{p}) = d_0 + \frac{d_1}{\mathrm{T}} - d_2 \ln \mathrm{T} \qquad (3)$$

where d_i are parameters with physical meaning and generally pressure dependent. The model, as given by the combination of eqs. (2) and (3), is the simplest one capable of rationalizing all known types of binary L-L phase separation in the T,p,x space. Phase separation is dictated by the attainability of a certain critical value for χ (in other words, for the excess molar Gibbs energy of the mixture), which is dependent on the ratio, r, between the segment-number of solute (1) to solvent (2). This ratio is usually closely related to the ratio of molar volumes ($r \sim V_1/V_2$). Parameters of eq. (3) and r (which determines the temperature independent segment fraction, φ) are obtained by a least-squares fitting to sets of experimental cloud-point data. Once they are set, many other thermodynamic properties of the mixture are also. For instance, one should recall that the excess molar enthalpy is intimately related to the temperature derivative of the mixture's Gibbs energy, while the excess volume is determined from its pressure derivative. We showed (5a) that this model is totally compatible with the so-called Prigogine and Defay equation, which, under some restrictive assumptions (17), establishes a Clapeyron-type relationship between, on the one hand, the pressure dependence of the critical temperature and, on the other, the ratio between the excess volume and enthalpy,

$$\left(\frac{\mathrm{d}T}{\mathrm{d}p}\right)_{\mathrm{c}} \cong T_{\mathrm{c}}(p)\frac{\Delta V^{\mathrm{E}}(T_{\mathrm{c}}(p),x)}{\Delta H^{\mathrm{E}}(T_{\mathrm{c}}(p),x)} \qquad (4)$$

Figure 4 depicts the model results for the binodal and spinodal at a nominal pressure of 1 bar in comparison with experimental cloud-point data for three systems (A-C) and Table III the corresponding parameters ($d_2 = 0$) of the model for the same systems.

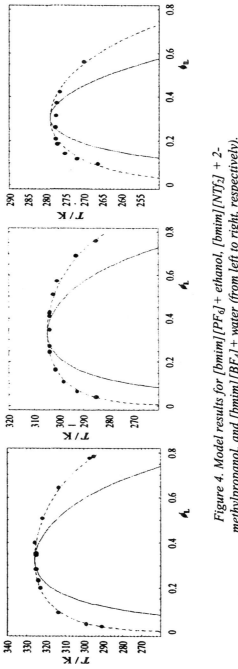

Figure 4. Model results for [bmim][PF₆] + ethanol, [bmim][NTf₂] + 2-methylpropanol, and [bmim][BF₄] + water (from left to right, respectively). Spinodal (full line) and binodal (dashed line). Filled circles are experimental points.

Table III: Model results for A = [bmim][PF₆]+ ethanol, B = [bmim][NTf₂] + 2-methylpropanol, and C = [bmim][BF₄]+ water

	A	B	C
	r = 4.6	r = 4.2	r = 5.3
Rd_1/J mol^{-1}	10397	13119	6263
Rd_0/J mol^{-1} K^{-1}	-22.97	-33.85	-13.87
T_c/K	325.9	304.7	279.1
$\varphi_{1,c}$	0.318	0.328	0.303
x_1^{max} (a)	0.318	0.328	0.303
H^E_{max}/J mol^{-1}	+4830	+5920	+3033
$V^E_{max} \cdot 10^6$/m^3 mol^{-1}	-3.0	-0.50	+0.22

(a) It represents the mole fraction of IL at the extremum of the excess properties.

Note that, in the case of the aqueous system, the value of the effective, optimized r is about two times smaller than that which would be predicted *a priori* by the mere analysis of volumes ratio. This suggests that water molecules tend to be associated in clusters which, on average, are well represented by dimers.

As expected for mixtures which phase separate on cooling (UCST), all H^E's are positive. For those mixtures involving alcohols (where dT/dp is negative – see Table II) the model's results lead to a negative V^E, which becomes positive if water is the solvent. The particular case of [bmim][BF$_4$] + water is analyzed in some detail in another contribution to this book (*18*) but here we can anticipate that the predicted value of +0.22 cm^{3}mol^{-1} for the excess volume at its extremum (see Table III) was very recently checked against experimental results. Preliminary experimental V^E's for this system at a nominal pressure of 1 bar are plotted in Figure 5 at 333.15 K and 277.75 K (very close to the critical temperature). Taking into account that only phase equilibria information was used as input data to the model, the agreement between predicted and experimental V^E is good. Cerdeiriña et al. (*18*) report experimental H^E of this same system. Again, it is rewarding to verify that the model's estimation agrees reasonably well with the experimental evidence.

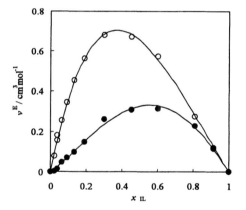

Figure 5. Excess volumes of [bmim][BF$_4$] + water at 333.15 (open circles) K and 277.75 K (filled circles).

Judging by the values reported for H^E and V^E for the systems analyzed in Table III, it seems that, while the mixing of an IL with a given solvent is generally accompanied by a significant enthalpic effect, the corresponding volumetric effect is pratically absent (quasi-ideal volumetric mixing).

Speeds of sound and densities of ILs

Speed of sound, u, and mass density, ρ, data can, in principle, be combined in order to determine other thermodynamic properties according to the following well-known thermodynamic relations:

$$\kappa_S = \frac{1}{u^2 \rho}; \qquad \alpha_p = -\frac{1}{\rho}\left(\frac{\partial \rho}{\partial T}\right)_p; \qquad \kappa_T = \frac{1}{\rho}\left(\frac{\partial \rho}{\partial p}\right)_T; \qquad c_p = \frac{T}{\rho}\frac{\alpha_p^2}{\kappa_T - \kappa_S} \qquad (5)$$

We have just determined the speed of propagation of ultrasound waves (1 MHz) in three ILs, namely, [bmim][BF$_4$], [bmim][PF$_6$], and [bmim][NTf$_2$] in broad temperature and pressure ranges (1 < p/bar < 1500; 283.15 < T/K < 323.15). To the best of our knowledge these constitute the first data of sound-speed in ILs. Densities have been determined for the same ILs – except for [bmim][PF$_6$] where literature data were found (*19*) - in the following temperature and pressure ranges: 1 < p/bar < 600; 298.15 < T/K < 333.15. Table IV compares u, ρ, V_m, κ_s, κ_T, and c_p/c_v for the three ILs at 298.15 K and two sample pressures (1 bar and 500 bar). Figure 6 plots the isobaric expansion coefficients versus temperature at a nominal pressure of 1 bar.

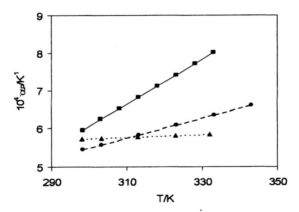

Figure 6. Isobaric expansion coefficient versus temperature at 1 bar of [bmim][BF$_4$] (▲), [bmim][PF$_6$] (●), and [bmim][NTf$_2$] (■).

As the anion increases in size ([BF$_4$]$^-$ < [PF$_6$]$^-$ < [NTf$_2$]$^-$) all the above-mentioned quantities increase accordingly, except the sound-speed which shows the opposite trend. At first glance, this fact may seem counterintuitive as one tends to associate greater density with greater sound-speed. This line of thought mainly arises from comparisons (in density and sound-speed) made between gas-phase and condensed-phase. It is true that gases show lower densities as well as lower speeds of propagation of sound waves than condensed phases, but this truism has little to do with density (see first relation in eqs. (5)). The typical ratio between densities of liquid to gas (far from the critical point) is a figure ranging from, say, 500 to 1,000. But the ratio (gas to liquid) of compressibilities is c.a. 100,000. Therefore, compressibility is the factor which controls the sound-speed. This statement is also well illustrated by analyzing the variation of these three quantities as pressure is applied. Note that, as pressure rises, density increases, compressibility decreases, and the sound-speed increases. In the particular case of these ILs, both compressibilities and densities increase as the anion gets larger and, thus, both quantities contribute to lowering the speed-sound (see first relation in eqs. (5)). It is also interesting to note that, in these ILs (differing in the anion), if the mass density increases the molar volume also does so. Thus, the distinct molar masses are playing here the most important role.

Table IV. Several thermodynamic properties* of selected [bmim]⁻ -based ILs (effect of the anion) at 298.15 K and two pressures, 1 and 500 bar

	[bmim][BF₄]		[bmim][PF₆]		[bmim][NTf₂]	
	1 bar	500 bar	1 bar	500 bar	1 bar	500 bar
u/ms^{-1}	1556.9	1686.3	1421.9	1543.7	1230.3	1381.6
ρ/Kgm^{-3}	1205.02	1226.04	1360.30	1383.81	1432.23	1465.86
V_m/cm^{3}mol^{-1}	187.57	184.36	208.91	205.36	292.81	286.09
$10^5 \kappa_S$/bar^{-1}	3.42	2.87	3.64	3.04	4.61	3.57
$10^5 \kappa_T$/bar^{-1}	3.65	3.29	3.99	3.25	5.41	4.02
$c_\mathrm{p}/c_\mathrm{v}$	1.07	1.14	1.09	1.07	1.17	1.13

* densities and quantities derived from density of [bmim][PF₆] are taken from Ref. (*19*).

Generally, one can state that sound-speed values of ILs are not too dissimilar to those typically found in conventional solvents, tending to fall in the high-value side of the range. In contrast, both compressibilities and expansivities (see Table IV and Fig. 6) at about room temperature and atmospheric pressure are *c.a.* three times smaller than values typically found in those conventional solvents. This is probably a consequence of the fact that although critical temperatures are unknown for ILs they are certainly much higher (were it possible to reach the liquid-gas critical point by avoiding thermal decomposition of the IL) than those of molecular solvents. Therefore, in the case of ILs, room-temperature conditions certainly correspond to very low reduced temperatures. It is also well known that both compressibility and expansivity decrease drastically along the orthobaric line from an infinite value at the critical point to very modest values as temperature (and reduced temperature) decreases.

According to Figure 6 expansivities increase with increasing temperature at constant pressure. This behavior contradicts that which has been reported both in other experiments (19) as well as in simulations (20). We have decided to reanalyze the raw data for density from the literature (19) available for [bmim][PF$_6$] and obtained the curve for α_p = f(T) that is depicted in Fig. 6. From this reanalysis plus our own data for the other two ILs we conclude that expansivities of these ILs increase as temperature rises.

It is already well-known that ILs are easy to metastabilize, meaning that, on cooling, they may remain liquid below their equilibrium melting line (supercooling). A liquid is said to be supercooled when it is found as such either at temperatures lower than those of the melting line or at pressures above it (for positively sloped melting lines). For [bmim][PF$_6$] we have several times witnessed solidification in our isothermal runs intended to determine the sound-speed as a function of rising pressure. Since a crystalline phase cannot be superheated (21), whenever solidification occurred we decided to lower the pressure slowly until the liquid state was again obtained. This way, we were able to detect several points of the melting line of [bmim][PF$_6$] since the difference between the sound-speed in this IL between the two states (solid and liquid) is significant ($u_{sol} - u_{liq} \sim 1000$ ms^{-1}). Within the experimental uncertainty of our results ($\pm$ 0.1 K; $\pm$ 10 bar) for the p-T melting line, it is a straight line between the coordinates (285.15 K; 1 bar) and (~305 K; ~1000 bar), evolving with a slope dp/dT = 48.7 barK^{-1}. The first coordinate corresponds to our visual determination of melting at atmospheric pressure obtained in a sealed glass vessel by multiple cycles of slow temperature change. It agrees reasonably well with that reported for water-free [bmim][PF$_6$] in Ref. (1d) (283.15 K), that of Ref. (22) (283.51 K) and disagrees with that claimed in Ref. (13i) (276.43 K) or that reported for water-equilibrated [bmim][PF$_6$] (1d) (277.15 K). By identifying the enthalpy of fusion, it is therefore possible to determine the volume change on solidification. Solidification in [bmim][PF$_6$] is accompanied by a contraction of about 7 %. This figure is highly dependent on the standartization of the enthalpy of fusion. We note that literature values do not agree. While Magee et al. (22) report a value of 19.6 kJmol^{-1}, Domanska and Marciniak (13i) determined a value of 9.21 kJmol^{-1}. In our calculations we have chosen the first value as their melting temperature agrees reasonably well with ours.

Theoretically, it is possible to determine heat capacities by using the last relation expressed in eq. (5). The overall uncertainty associated with this type of calculation lies significantly on the magnitude of the difference ($\kappa_T - \kappa_s$). We note that $\kappa_T / \kappa_s = c_p/c_v$. In a previous work (12) we successfully determined the heat capacity of 2-propanone (acetone) as a function of temperature and pressure using density and sound-speed data. Our results showed that the ratio (κ_T / κ_S) in acetone varies, irrespective of temperature, linearly from 1.4 to 1.3 as pressure shifts from 1 to 600 bar. In other words, κ_T is greater than κ_S by 30 to 40 %. Taking into account that the accuracy of both κ_T and κ_S is to the order of a few percent, the calculation of c_p based on eq. (5) can still be legitimated, but the difference in the denominator largely contributes to the overall uncertainty of c_p. It occurs that in ILs one does not find such a favorable situation. For instance, in the case of [bmim][BF$_4$], the ratio (κ_T / κ_S) varies linearly from *c.a.* 1.07 to 1.15 as pressure shifts from 1 to 600 bar (almost irrespective of temperature). So, calculations should not be validated at atmospheric or moderately low pressure, but, paradoxically they may have significance at high pressure. In a sense, this is good, for the main advantage of using eq. (5) as a tool for determining heat capacities in comparison with a direct calorimetric determination is that the latter is seldom performed when pressure is a variable of interest.

We conclude this section with a note on the "ideal" volumetric behavior of ILs. During the progress of this research project, which has involved 1-R$_n$-3-methylimidazolium-based ILs, we have accumulated a considerable amount of density data as a function of temperature (and pressure). Figure 7 plots the

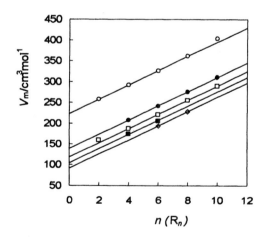

Figure 7. Comparison of experimental molar volumes of 1-R$_n$-3-methylimidazolium-based ILs with those predicted. See text and Table V. Series of [Cl]$^-$ (◇),[NO$_3$] (■), [BF$_4$] (□),[PF$_6$] (●), and [NTf$_2$] (○).

molar volumes of this class of ILs (for several different anions) versus the number, n, of carbon atoms in the alkyl chain, R_n, at 298.15 K and atmospheric pressure. Along with others ($1d,19,23$) we note an impressive degree of linearity in the plots. Moreover, a master slope, ($\partial V_m/\partial(2n)$), for the variation of molar volume per addition of two carbon atoms, ($\partial V_m/\partial(2n)$) = (34.4 ± 0.5 cm^3mol^{-1}), is obtained irrespective of the anion considered ([Cl⁻], [NO₃⁻], [BF₄⁻], [PF₆⁻], and [NTf₂⁻]).

Were the accuracies of the experimental density data available better and more consistent along each series, it is our contention that the uncertainty in the master slope would significantly decrease. That linearity prompted us to try to anchor the effective volume occupied by one type of anion in one mol of IL (one mol of anion and one mol of cation) from which one can establish the effective volume ("molar size") occupied by all the other anions and cations. This can be used as a predictive tool for the estimation of molar volumes of ILs.

Tentatively, we anchored the "size" of the PF₆⁻ anion by using 1.73 Å for the P-F bond length and 1.35 Å for the van der Waals radius of the fluorine ion ($13k$). In other words, we set the size of this anion as an equivalent hard sphere of radius equal to 3.08 Å. This value leads to an effective volume of 73.71 cm^3mol^{-1}. Differences between the effective volumes occupied by different anions are merely taken from the differences between the corresponding straight lines depicted in Fig. 7. For instance, the difference between molar volumes of PF₆⁻-based ILs and BF₄⁻-based ones is 20.29 cm^3mol^{-1}, irrespective of the length of the alkyl chain in the imidazolium. This sets the effective volume of BF₄⁻ as 53.42 cm^3mol^{-1}; and so forth for the remaining anions. We tested how theoretically sound this value would be for the tetrafluoroborate anion by reversing the calculation. That effective molar volume corresponds to an effective radius of the equivalent hard sphere of BF₄⁻ of 2.77 Å, or, in other words, it establishes the B-F bond length as 1.42 Å to be compared with the literature value ($13k$) of 1.4 Å.

The cations' sizes were established starting at n = 0 (hydrogen), i.e., at the y-intercept of the plots of Fig. 7. Then, for longer alkyl chains, one applies the recipe of adding 34.4 cm^3mol^{-1} per two additional carbons. The overall results are reported in Table V, where the figures in boldface correspond to the individual contributions of the anions and the cations (n=0 through n=12) and the internal matrix is obtained by the mere sum of the volumes of the anion-cation pair. Experimental values, when they are available, appear inside brackets

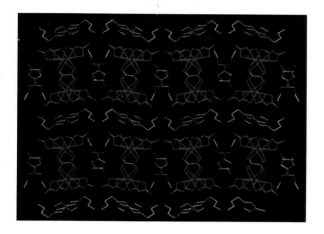

Figure 3.1. Illustration of the cation-anion arrangement in 1 shows the cations surrounding the anion columns. Hydrogen atoms have been omitted for clarity. (Reproduced from *Inorg. Chem.* **2004**, *43(8)*, 2503-2514. Copyright 2004 American Chemical Society.)

Figure 3.2. Illustration of the cation-anion arrangement in 2 shows the cations and solvate ionic liquid surrounding the anion columns. Hydrogen atoms have been omitted for clarity. (Reproduced from *Inorg. Chem.* **2004**, *43(8)*, 2503-2514. Copyright 2004 American Chemical Society.)

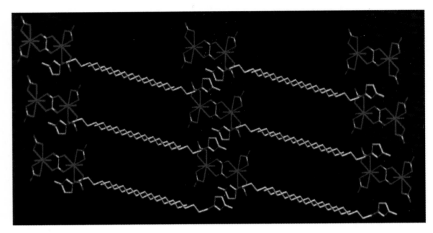

Figure 3.3. Illustration of the cation-anion arrangement in 3 showing the anions associated with cationic head groups and channels between the alkyl chains regions. The MeCN molecules have been omitted for clarity.
(Reproduced from *Inorg. Chem.* **2004**, *43(8)*, 2503-2514. Copyright 2004 American Chemical Society.)

Figure 21.1. *Non-intrusive microcell for sound-speed measurements in ILs. The ultrasonic transducers are not in direct contact with the liquid under study.*

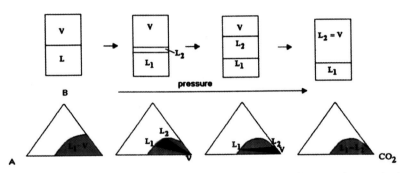

Figure 23.1. *Correspondence between the succession of states observed when carbon dioxide is added isothermally to a mixture of ionic liquid (A) and an alcohol or alcohol + water (B).*

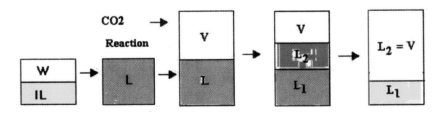

Figure 23.3. *Water + [C_4mim][PF_6] mixtures change the number of phases with successive addition of ethanol (first transition) and carbon dioxide.*

Table V. Molar "sizes" of cations and anions and corresponding predicted molar volumes of ILs (see text). Experimental values appear inside brackets

		BF$_4$	PF$_6$	NTf$_2$	NO$_3$	Cl
		53.42	73.71	158.66	39.13	25.86
R$_0$	64.82	118.24	138.53	223.48	103.95	90.68
R$_2$	99.20	152.62 (159.14[a])	172.91	257.86 (257.61)	138.33	125.06
R$_4$	133.58	187.00 (187.62[a])	207.29 (207.82)	292.24 (292.11)	172.71 (174.44[a])	159.44
R$_6$	167.96	221.38 (221.07[a])	241.67 (241.45)	326.62 (326.21)	207.09 (205.36[a])	193.82 (194.07[b])
R$_8$	202.34	255.76 (255.48[a])	276.05 (275.39)	361.00 (361.81)	241.47	228.20 (227.95[a])
R$_{10}$	236.72	290.14 (290.11[a])	310.43 (310.79)	395.38 (403.32)	275.85	262.58
R$_{12}$	271.10	324.52	344.81	429.76	310.23	296.96

(a) from ref. (23a); (b) from ref. (23b).

Note that some predictions reported in Table V may refer to metastable liquid conditions. Considering the crudeness of the calculation, the agreement between experimental and predicted values is excellent (better than 1 %, and often much better) in all but two cases. In these cases, the experimental IL density is lower than predicted, and this can be attributed to the presence of a non-negligible amount of water in the IL samples.

The common master slope shown in Fig. 7 seems to constitute a fingerprint of ideal behavior in regard to volumetric properties of ILs. The volume of ILs increases by almost exactly the same amount as the addition of units in the alkyl chain in the methylimidazolium cation proceeds (irrespective of the interactions with totally different anions differing in size, shape, and chemical structure). This suggests that binary mixtures constituted by pairs of ILs selected from Table V will be formed with almost null variation of volume (almost null excess molar volume). This is, in fact, what is experimentally observed. Among the several possible combinations that Table V suggests, we selected some cases where binary mixtures of ILs have either a commom cation or a commom anion. This is a particulary interesting situation because, in these cases, any observed excess volume is a consequence of the interaction of two distinct anions (or cations) in a constant cation's (or anion's) field. Note that in the same spirit as above where we hypothetically divided the entire volume of the IL ($V_{IL(i)}^{\circ}$) into that effectively occupied by the anion ($V_{A(j)}^{*}$) and that of the cation ($V_{C(k)}^{*}$), one obtains (e.g., in the case of a common cation)

$$V^E = V_{mix} - x_1 V_{IL(1)}{}^o - x_2 V_{IL(2)}{}^o = V_{mix} - x_1 (V_C{}^* + V_{A(1)}{}^*) - x_2 (V_C{}^* + V_{A(2)}{}^*) \quad (6)$$

where V^E and V_{mix} stand for the excess molar volume and molar volume of the binary mixture, respectively. After a straightforward rearrangement and defining the effective volume occupied by the mixture of anions in the binary mixture of ILs as

$$V_{A(1)+A(2)}{}^* = V_{mix} - V_C{}^* \quad (7)$$

Then,

$$V^E = V_{A(1)+A(2)}{}^* - x_1 V_{A(1)}{}^* - x_2 V_{A(2)}{}^* \quad (8)$$

Such an example is realized in Fig. 8 for the $[bmim]^+([NTf_2]^- + [BF_4]^-)$ ionic liquid mixture.

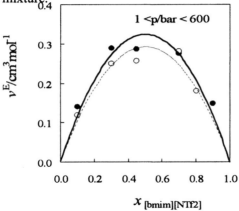

Figure 8. Excess molar volume of the $[bmim]^+([NTf_2]^- + [BF_4]^-)$ ionic liquid mixture. Full line and filled circles at 303.15 K; dashed line and open circles at 333.15 K (irrespective of the pressure applied).

As expected, the observed expansion corresponds solely to about 0.1 % of the molar volume of the equimolar mixture and it is impressively independent of pressure (up to 600 bar) to an uncertainty of $\pm$ 0.01 $cm^3 mol^{-1}$. We are not aware of any other binary mixture in which no shift in its excess volume upon pressurization up to several tens of megapascal was detected.

Table VI shows a similar situation but now the anion is maintained identical in both ILs ($[(R_n + R_k)mim]^+[NTf_2]^-$).

Table VI: Equimolar excess molar volumes at 298.15 K for ([(R$_n$ + R$_k$)mim]$^+$[NTf$_2$]$^-$) binary mixtures of ILs.

	R$_2$ + R$_{10}$	R$_4$ + R$_{10}$	R$_6$ + R$_{10}$	R$_8$ + R$_{10}$	R$_4$ + R$_8$
V^E ½ (cm^3 mol^{-1})	0.260	0.220	0.110	0.084	0.139

Again, the observed expansion due to the interaction of two distinct cations is very mild (less than 0.1 % of the molar volume of the mixtures) and, interestingly, the effects seem to be additive. Note the trend in the variation of the values of V^E for the equimolar mixtures as one moves from the situation n=2 + k=10 to that where n=8 + k=10. In spite of the fact that some selected binary mixtures of ILs can phase separate, even in the event of having a common ion (24), the corresponding volumetric behavior is quasi-ideal.

Acknowledgments: We are indebted to Prof. Kenneth R. Seddon for his continuous support and fruitful discussions. Work supported by *FC&T*, Portugal, contract # POCTI/EQU/35437/00.

References

1. (a) Welton, T. *Chem. Rev.* **1999**, *99*, 2071; (b) Holbrey, J. D.; Seddon, K. R. *Clean Prod. Process.* **1999**, *1*, 223; (c) Brennecke, J. F.; Maginn, E. J. *AIChE J.* **2001**, *47*, 2384; (d) Huddleston, J. G.; Visser, A. E.; Reichert, W. M.; Willauer, H. D.; Broker, G. A.; Rogers, R. D. *Green Chem.* **2001**, *3*, 156; (e) Seddon, K. R.; Stark, A.; Torres, M.-J. *Pure Appl. Chem.* **2000**, *72*, 2275; (f) Anthony, J. L.; Maginn, E. J.; Brennecke, J. *J. Phys. Chem. B* **2002**, *106*, 7315; (g) Domanska, U.; Bogel-Lukasik, E.; Bogel-Lukasik, R. *J. Phys. Chem. B* **2003**, *107*, 1858.
2. Visser, A. E.; Swatloski, R. P.; Reichert, W. M.; Griffin, S. T.; Rogers, R. D. *Ind. Eng. Chem. Res.* **2000**, *39*, 3596.
3. Najdanovic-Visak, V.; Esperança, J. M. S. S. ; Rebelo, L. P. N.; Nunes da Ponte, M.; Guedes, H. J. R.; Seddon, K. R.; Szydlowski, J. *Phys. Chem. Chem. Phys.* **2002**, *4*, 1701.
4. (a) Fadeev, A. G.; Megher, M. M. *Chem. Commun.* **2001**, 295; (b) Freemantle, M. *Chem. Eng. News* **2001**, *April 2*, 57.
5. (a) Rebelo, L. P. N. *Phys.Chem.Chem.Phys.*, **1999**, *1*, 4277; (b) de Sousa, H. C.; Rebelo, L. P. N.; *J. Polym. Sci. B: Polym. Phys.* **2000**, *38*, 632.

6. (a) Bigeleisen, J. *J.Chem.Phys.*, **1961**, *34*, 1485; (b) Jancsó, G.; Rebelo, L. P. N.; Van Hook, W. A. *Chem. Rev.* **1993**, *93*, 2645; (c) Van Hook, W. A.; Rebelo, L. P. N.; Wolfsberg, M. *J. Phys. Chem. A* **2001**, *105*, 9284.

7. Trusler, J.P.M., *Physical Acoustics and Metrology of Fluids*, Adam Hilger, Bristol, 1991

8. Douhéret, G.; Davis, M.I.; Reis, J.C.R.; Blandamer, M.J. *Chem. Phys. Chem.* **2001**, *2*, 148.

9. (a) Gordon, C. M.; Holbrey, J. D.; Kennedy, R.; Seddon, K. R. *J. Mater. Chem.* **1998**, *8*, 2627; (b) Bonhôte, P.; Dias. A.-P.; Armand, M.; Papageorgiou, N.; Kalyanasundaram, K.; Gratzel, M. *Inorg. Chem.* **1996**, *35*, 1168.

10. de Sousa, H. C.; Rebelo, L. P. N. *J. Chem. Thermodyn.* **2000**, *32*, 355.

11. Rebelo, L. P. N.; Visak, Z. P.; de Sousa, H. C.; Szydlowski, J.; Gomes de Azevedo, R.; Ramos, A. M.; Najdanovic-Visak, V.; Nunes da Ponte, M.; Klein, J. *Macromolecules* **2002**, *35*, 1887.

12. Gomes de Azevedo, R.; Szydlowski, J.; Pires, P.F.; Esperança. J.M.S.S.; Guedes, H.J.R.; Rebelo, L.P.N. *J. Chem. Thermodyn.* **2003**, *36*, 211.

13. (a) Najdanovic-Visak, V.; Esperança, J. M. S. S. ; Rebelo, L. P. N.; Nunes da Ponte, M.; Guedes, H. J. R.; Seddon, K. R.; de Sousa, H.C.; Szydlowski, J. *J. Phys. Chem B.* **2003**, *107*, 12797; (b) Swatloski, R. P.; Visser, A. E.; Reichhert, W. M.; Broker, G. A.; Farina, L. M.; Holbrey, J. D.; Rogers, R. D. *Chem. Commun.* **2001**, 2070; (c) Swatloski, R. P.; Visser, A. E.; Reichhert, W. M.; Broker, G. A.; Farina, L. M.; Holbrey, J. D.; Rogers, R. D. *Green Chem.* **2002**, *4*, 81; (d) Anthony, J. L.; Maginn, E. J.; Brennecke, J. F. *J. Phys. Chem. B* **2001**, *105*, 10942; (e) Marsh, K.N.; Deev, A.; Wu, A. C.-T.; Tran, E.; Klamt, A. *Korean J. Chem. Eng.* **2002**, *19*, 357; (f) Wu, C.-T.; Marsh, K.N.; Deev, A.V.; Boxall, J.A. *J. Chem. Eng. Data* **2003**, *48*, 486; (g) Heintz, A.; Lehmann, J. K.; Wertz, C. *J. Chem. Eng. Data* **2003**, *48*, 472, (h) Wong, D.S.H.; Chen, J.P.; Chang, J.M.; Chou, C.H. *Fluid Phase Equilib.* **2002**, *194-197*, 1089; (i) Domanska, U.; Marciniak, A. *J. Chem. Eng. Data* **2003**, *48*, 451; (j) Domanska, U.; Marciniak, A *J. Phys. Chem. B* **2004**, *108*, 2376; (k) Wagner, M; Stanga, O.; Schröer, W. *Phys. Chem. Chem. Phys.* **2003**, *5*, 3943.

14. Jancsó, G. *KFKI Közl.*, **1966**, *14*, 219.

15. Visak, Z. P.; Rebelo, L. P. N.; Szydlowski, J. *J. Phys. Chem. B* **2003**, *107*, 9837.

16. Qian, C.; Mumby, S.J.; Eichinger, B.E. *J. Polym. Sci.: Polym. Phys.* **1991**, *29*, 635.

17. Rebelo, L. P. N.; Najdanovic-Visak, V.; Visak, Z. P.; Nunes da Ponte, M.; Troncoso, J.; Cerdeiriña, C. C.; Romani, L. *Phys.Chem.Chem.Phys.* **2002**, *4*, 2251.

18. Cerdeiriña et al., (this same book).
19. Gu, Z.; Brennecke, J. *J. Chem. Eng. Data* **2002**, *47*, 339.
20. Morrow, T. I.; Maginn, E. J. *J. Phys. Chem. B* **2002**, *106*, 12807
21. Debenedetti, P.G. *Metastable Liquids*, Princeton Univ. Press, Princeton, N.J., 1996.
22. Magee, J.W. *in Book of Abstracts 17th IUPAC Conf. Chem. Thermodyn.*, Rostock, Germany, 2002, pp. 306.
23. (a) Seddon, K.R.; Stark, A.; Torres, M.-J. *in Clean Solvents: Alternative Media for Chemical Reactions and Processing*, Eds. M. Abraham and L. Moens, ACS Symp. Ser., Vol. 819, American Chemical Society, Washington D.C., 2002, pp. 34; (b) Torres, M.-J. *Ph.D. Thesis*, The Queen's University of Belfast, Belfast, 2001; (c) Suarez, P.A.Z.; Einloft, S.; Dullius, J.E.L.; de Souza, R.F.; Dupont, J. *J. Chim. Phys.* **1998**, *95*, 1626; (d) Bonhôte, P.; Dias, A.-P.; Papageorgiou, N.; Kalyanasundaram, K.; Gratzel, M. *Inorg. Chem.* **1996**, *35*, 1168.
24. Seddon, K.R. – *private communication*, 2003. For instance, we have also witnessed in our labs that mixtures of two ILs where a common anion is present (chloride) but the cations differ (dialkylimidazolium versus trihexyl(tetradecyl)phosphonium), such as $[C_4mim][Cl]$ + $[(C_6H_{13})_3P(C_{14}H_{29})][Cl]$ as well as $[C_5mim][Cl]$ + $[(C_6H_{13})_3P(C_{14}H_{29})][Cl]$, phase split in a broad temperature range which includes room temperature. Nonetheless, excess volumes of similar but one-phase mixtures (where $[C_4mim]^+$ or $[C_5mim]^+$ have been replaced by $[C_6mim]^+$ or $[C_8mim]^+$) are very small.

Chapter 22

Phase Equilibria with Gases and Liquids of 1-*n*-Butyl-3-methylimidazolium Bis(trifluoromethylsulfonyl)imide

Jacob M. Crosthwaite, Laurie J. Ropel, Jennifer L. Anthony, Sudhir N. V. K. Aki, Edward J. Maginn, and Joan F. Brennecke

Department of Chemical and Biomolecular Engineering, University of Notre Dame, South Bend, IN 46556

The solubility of carbon dioxide in 1-*n*-butyl-3-methylimidazolium bis(trifluoromethylsulfonyl)imide ([bmim][Tf$_2$N]) at 25 ^{0}C is reported and compared with CO$_2$ solubility in 1-*n*-butyl-3-methylimidazolium hexafluorophosphate and tetrafluoroborate ([bmim][PF$_6$] and [bmim][BF$_4$]). In addition, we report liquid-liquid equilibrium data for [bmim][Tf$_2$N] with l-butanol and 1-hexanol. By comparison with literature data, we show the influence of anion and alkyl chain length on the imidazolium ring on phase behavior with alcohols. Finally, we present an estimate for the octanol/water partition coefficient, which is a measure of potential bioaccumulation, for [bmim][Tf$_2$N].

Here we present preliminary results for the solubility of carbon dioxide in 1-*n*-butyl-3-methylimidazolium bis(trifluoromethylsulfonyl)imide([bmim] [Tf$_2$N]). Previously, we have measured the solubilities of a wide variety of gases in 1-*n*-butyl-3-methylimidazolium hexafluorophosphate ([bmim][PF$_6$]) (1), as well as the solubility of carbon dioxide and methane in 1-*n*-butyl-3-methylimidazolium tetrafluoroborate ([bmim][BF$_4$]) (2). The solubility of these two gases in [bmim][PF$_6$] and [bmim][BF$_4$J are relatively similar. The goal of the current work with [bmim][Tf$_2$N] is to explore further whether there is any influence of the nature of the anion on gas solubilities. We know of no other published values of the solubility of gases in [bmim][Tf$_2$N]. We are particularly interested in understanding the solubility of gases in ionic liquids due to the potential of using these non-volatile solvents for gas separations (3-5).

Second, we present the liquid-liquid equilibrium (LLE) of 1-butanol and 1-hexanol with [bmim][Tf$_2$N]. These results are part of a systematic study that we are conducting of the LLE of alcohols with imidazolium-based ILs as a means of understanding the characteristics of the IL (e.g., cation, choice of anion, substituents on the cation) that are important in determining the phase behavior. LLE is of practical importance for understanding the solubility of reactants and products in ILs, as well as the amount of IL that will contaminate organic or aqueous phases that are brought into contact with the IL. While a number of researchers have presented LLE results for various dialkylimidazolium salts with alcohols (6-9), we know of only one study of LLE of any dialkylimidazolium Tf$_2$N compounds with alcohols (10). All of these systems show upper critical solution temperature (UCST) behavior; i.e. the mutual solubilities increase with increasing temperature.

Finally, we report an estimate for the octanol/water partition coefficient, K$_{ow}$, of [bmim][Tf$_2$N]. K$_{ow}$ is defined as the ratio of the concentration of a solute in the octanol phase to the concentration of the same solute in the water phase in a two phase mixture of octanol, water, and solute. This definition is valid when the solute concentration is very dilute. K$_{ow}$ values provide estimates of the partitioning of a solute between water and the lipid layers of aquatic organisms, as well as the distribution of a solute between water and sediments (11). Thus, K$_{ow}$ gives a first estimate of bioaccumulation potential. Moreover, K$_{ow}$ values have been correlated with the toxicity to aquatic organisms (11). Partitioning of alkylmethylimidazolium chlorides, where the alkyl chain ranged from butyl to dodecyl, between octanol and water have been reported (9). However, these measurements were made for highly concentrated solutions and cannot be compared directly with the dilute K$_{ow}$ value reported here. In addition, Choua et al. (12) reported K$_{ow}$ values for [bmim][PF$_6$] and ethyl-3-methylimdazolium hexafluorophosphate ([emim][PF$_6$]) that ranged from 0.005 to 0.020 using a shake-flask method without centrifugation. This method can lead to inaccurate results if the emulsion is not fully separated.

Experimental

A gravimetric microbalance (IGA 003, Hiden Analytical) was used to measure the low pressure (less than 14 bar) gas solubilities reported in this work. This microbalance and the technique used to measure gas solubilities in ILs has been described in detail elsewhere (1,13). Briefly, a small IL sample is placed on the balance and the system is thoroughly evacuated to remove any volatile impurities, including water. The gas is introduced and the gas solubility is determined from the mass uptake.

The high pressure (up to about 65 bar) CO_2 solubility in [bmim][Tf$_2$N] was measured using a static high-pressure apparatus that has been described previously (14). Known amounts of CO_2 are metered into a cell containing IL. Using the known volume of the vapor space and an equation of state to determine the number of moles in the vapor, the solubility of CO_2 in the IL is determined by the difference of these two amounts of substance.

The liquid phase behavior of [bmim][Tf$_2$N] with 1-butanol and 1-hexanol was measured using a cloud point technique described previously (2).

The slow-stirring method was used for determining the octanol/water partition coefficient of [bmim][Tf$_2$N]. The experimental apparatus consisted of a sealed vial, equipped with a dip-tube to prevent contamination of the syringe needle when a sample of the lower phase is taken. The deionized water and 1-octanol were initially mixed together so they were in equilibrium prior to use. The pre-equilibrated deionized water was added to the 40 mL clear glass vial containing a Teflon coated magnetic stir bar. The lid consisted of a cap, septum, and a dip-tube that penetrates through the septum. This lid was placed on the vial so that the dip-tube extended below the surface of the water-rich phase. A known (dilute) concentration of [bmim][Tf$_2$N] was added to the pre-equilibrated 1-octanol separately. This solution was carefully added to the vial containing the water so that the solution did not emulsify. Then the vial was sealed. Five vials using different dilute octanol-[bmim][Tf$_2$N] "stock" solutions were made to determine the concentration dependence. The vials were stirred slowly to prevent emulsification and maintained at room temperature, which was 24 ± 2 °C. Three syringe samples were taken of the octanol-rich and water-rich phases over the course of ten days, at which time the concentrations in each phase had stabilized. Concentrations of [bmim][Tf$_2$N] in each phase were measured using UV-vis spectroscopy (Cary 3, Varian), which has a sensitivity of ± 0.01 absorption units.

For all the experiments, the [bmim][Tf$_2$N] ionic liquid was purchased from Covalent Associates. The 1-butanol and 1-hexanol for the LLE measurements were both 99+% anhydrous grade from Aldrich. The 99+% HPLC grade 1-octanol was also purchased from Sigma-Aldrich. The CO$_2$ was Coleman Instrument grade, obtained from Mittler Supply Company. All compounds were used without further purification. However, the IL was thoroughly dried at 80 °C under high vacuum and the water content was measured using Karl-Fisher titration (EM Science Aquastar V-200 Titrator) prior to use. Water content for the IL used in the high pressure gas solubility measurements was less than 500 ppm. For the LLE measurements, the IL water content was 460 ppm. In addition, the water content of the 1-butanol and 1-hexanol were 250 ppm and 590 ppm, respectively. In the low pressure gas solubility measurements with the gravimentric microbalance, the IL was dried further in situ with a 10^{-9} bar vacuum until the sample mass stabilized.

Results and Discussion

Gas Solubility. The solubility isotherm at 25 °C for CO$_2$ in [bmim][Tf$_2$N] is shown in Figure 1, along with our previously reported isotherms for CO$_2$ in [bmim][PF$_6$] (1) and [bmim][BF$_4$] (2), also at 25 °C. As can be seen in the graph, the solubility in [bmim][Tf$_2$N] is significantly larger than was seen with the other two ionic liquids. Changing the anion from [BF$_4$] to [PF$_6$] did not result in a significant change in affinity for CO$_2$, whereas the [Tf$_2$N] anion does increase this affinity.

A Henry's law constant (H) can be calculated from the isotherms as the inverse slope at low pressures (H = P/x) to yield the Henry's constant in units of pressure. At 25 °C, this Henry's constant is 33.0 $\pm$ 3 bar for CO$_2$ absorbed in [bmim][Tf$_2$N]. By comparison, the values for CO$_2$ in [bmim][PF$_6$] and [bmim][BF$_4$] are 53.4 $\pm$ 0.3 bar (1) and 56.5 $\pm$ 5 bar (2), respectively. Again, the higher solubility of CO$_2$ with the [Tf$_2$N] anion compared to either the [PF$_6$] or [BF$_4$] anions is seen by the smaller Henry's constant.

We have also measured the solubility of CO$_2$ in [bmim][Tf$_2$N] at higher pressures and the results are shown in Figure 2. The solubility measurements were repeated to ensure reproducibility and the results from both sets of data are shown in the figure. As expected, the solubility increases with an increase in pressure. For example, the solubility increased from 0.268 mole fraction at 11.4 bar to 0.717 mole fraction at the highest pressure measured (60.4 bar). Furthermore, we find excellent agreement between the low pressure data taken with the gravimetric microbalance and the higher pressure data taken with the static high pressure apparatus, as shown in the figure.

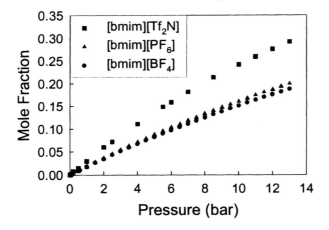

Figure 1: Solubility of carbon dioxide in [bmim][Tf$_2$N], [bmim][PF$_6$] (1) and [bmim][BF$_4$] (2) at 25 °C.

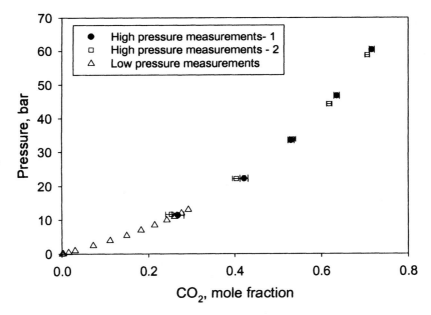

Figure 2: Solubility of carbon dioxide in [bmim][Tf$_2$N] at 25 °C using both the low pressure and high pressure apparatuses.

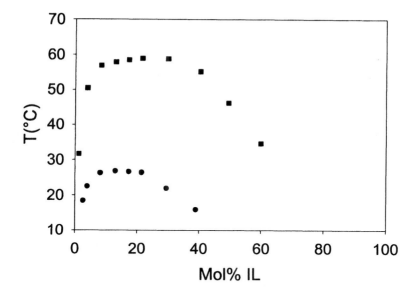

Figure 3: T-x diagram for [bmim][Tf₂N] with 1-butanol (•) and 1-hexanol (■).

Liquid Liquid Equilibrium. The liquid phase behavior of [bmim][Tf$_2$N] with 1-butanol and 1-hexanol is shown in Figure 3. From the figure, it is clear that these systems exhibit upper critical solution temperature behavior (UCST). In addition, the alcohol-rich phase of both systems contains a small amount of ionic liquid, while the ionic liquid-rich phase contains a large concentration of alcohol. The effect of alcohol chain length is clear by comparing the two curves: increasing the chain length of the alcohol (i.e., making the alcohol more aliphatic) increases the UCST of the system.

The effect of *cation* alkyl chain length is shown in Figure 4. By comparing our data for the 1-butanol/[bmim][Tf$_2$N] system with data for 1-butanol/1-ethyl-3-methylimidazolium bis(trifluoromethylsulfonyl)imide ([emim][Tf$_2$N]) from Heintz et al. (10), one observes that increasing the alkyl chain on the cation decreases the UCST of the system. This trend is likely due to increased interaction between the alkyl chains on the cation and alcohol via van der Waals forces. Also shown in Figure 4 is the dramatic effect that the anion has on the liquid phase behavior. A comparison of our data for the 1-butanol/[bmim][Tf$_2$N] system with data for the 1-butanol/[bmim][PF$_6$] system from Wu et al. (6) shows that changing the anion from [Tf$_2$N] to [PF$_6$] increases the UCST from about 26 °C to over 100 °C. Thus, the choice of anion can have a dramatic effect on LLE behavior.

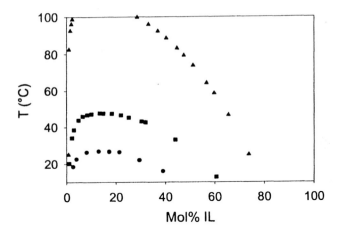

Figure 4: T-x diagram for [bmim][Tf$_2$N] (•) with 1-butanol. Comparison is made to [emim][Tf$_2$N] (■) from Heintz et al. (10) and [bmim][PF$_6$] (▲) from Wu et al. (6).

Octanol/Water Partition Coefficient. The octanol/water partition coefficient of [bmim][Tf$_2$N] at 24 ± 2 °C was determined to be between 0.11 and 0.62. Even at dilute concentrations (1.5 x 10^{-4} – 2.2 x 10^{-3} M in the aqueous phase) the K$_{ow}$ was concentration dependent (i.e., it was not at infinite dilution). These values show that [bmim][Tf$_2$N] preferentially partitions into water in the octanol/water two phase system. The following table compares the octanol/water partition coefficient of [bmim][Tf$_2$N] with those of some commonly used solvents. Clearly, K$_{ow}$ for [bmim][Tf$_2$N] is extremely small, similar to the values for polar organics that are completely miscible with water. This indicates that [bmim][Tf$_2$N] prefers water over octanol, even though [bmim][Tf$_2$N] exhibits immiscibility with water at room temperature (15,16). Moreover, the bioaccumulation potential is quite small.

Table 1: K$_{ow}$ of [bmim][Tf$_2$N] in comparison to common organic solvents.

Compound	K$_{ow}$
[bmim][Tf$_2$N]	0.11-0.62
methanol (17)	0.17
acetone (17)	0.575
benzene (18)	135
cyclohexane (18)	2754
n-hexane (18)	10000

Summary

The phase behavior of gases and liquids with 1-*n*-butyl-3-methylimidazolium bis(trifluoromethylsulfonyl)imide ([bmim][Tf$_2$N]) is strongly influenced by the anion. Specifically, the solubility of carbon dioxide in [bmim][Tf$_2$N] at 25 °C is greater than in either [bmim][PF$_6$] or [bmim][BF$_4$]. The upper critical solution temperature (UCST) of [bmim][Tf$_2$N] with 1-butanol is just above room temperature. This indicates a much greater affinity of 1-butanol for [bmim][Tf$_2$N] than for either [bmim][PF$_6$] or [bmim][BF$_4$], both of which have much higher UCSTs. Increasing the alcohol alkyl chain length or decreasing the cation alkyl chain length decreases mutual solubilities. Finally, we find that [bmim][Tf$_2$N] has low bioaccumulation potential, as indicated by the octanol/water partition coefficient.

Acknowledgements

Financial support from the National Science Foundation (grant CTS 99-87627), NOAA (grant NA16RP2892), and the Department of Education GAANN program (grant P200A010448-01).

References

1. Anthony, J. L.; Maginn, E. J.; Brennecke, J. F. *J. Phys. Chem. B* **2002**, *106*, 7315-7320.
2. Anthony, J. L.; Crosthwaite, J. M.; Hert, D. G.; Aki, S. N. V. K.; Maginn, E. J.; Brennecke, J. F. In Ionic Liquids as Green Solvents; Rogers, R. D., Seddon, K. R.; ACS Symposium Series 856; American Chemical Society: Washington, D.C.; 2002; pp. 110-120.
3. Brennecke, J. F.; Maginn, E. J. U.S. Patent 6,579,343, 2003.
4. Munson, C. L.; Boudreau, L. C.; Driver, M. S.; Schinski, W. L. U.S. Patent 6,339,182, 2002.
5. Munson, C. L.; Boudreau, L. C.; Driver, M. S.; Schinski, W. L. U.S. Patent 6,623,659, 2003.
6. Wu, C. T.; Marsh, K. N.; Deev, A. V.; Boxall, J. A. *J. Chem. Eng. Data* **2003**, *48*, 486-491.
7. Najdanovic-Visak, V.; Esperanca, J.; Rebelo, L. P. N.; da Ponte, M. N.; Guedes, H. J. R.; Seddon, K. R.; Szydlowski, J. *Phys. Chem. Chem. Phys.* **2002**, *4*, 1701-1703.
8. Wagner, M.; Stanga, O.; Schroer, W. *Phys. Chem. Chem. Phys.* **2003**, *5*, 3943-3950.

9. Domanska, U.; Bogel-Lukasik, E.; Bogel-Lukasik, R. *Chem.-Eur. J.* **2003**, *9*, 3033-3041.
10. Heintz, A.; Lehmann, J. K.; Wertz, C. *J. Chem. Eng. Data* **2003**, *48*, 472-474.
11. Allen, D. T.; Shonnard, D. R. *Green Engineering: Environmentally Conscious Design of Chemical Process*; 1st ed.; Prentice Hall PTR: Upper Saddle River, NJ, 2002.
12. Choua, C.-H.; Perng, F.-S.; Wong, D. S. H.; Su, W. C. In 15th Symposium on Thermophysical Properties; Boulder, CO, USA; 2003.
13. Anthony, J. L.; Maginn, E. J.; Brennecke, J. F. *J. Phys. Chem. B* **2001**, *105*, 10942-10949.
14. Blanchard, L. A.; Gu, Z. Y.; Brennecke, J. F. *J. Phys. Chem. B* **2001**, *105*, 2437-2444.
15. Crosthwaite, J. M.; Aki, S. N. V. K.; Maginn, E. J.; Brennecke, J. F. *J. Phys. Chem. B* **2004**, *108*, 5113-5119.
16. Bonhote, P.; Dias, A. P.; Papageorgiou, N.; Kalyanasundaram, K.; Gratzel, M. *Inorg. Chem.* **1996**, *35*, 1168-1178.
17. Syracuse Research Corporation. KOWWIN, Version 1.67, U.S. Environmental Protection Agency, 2000.
18. Ruelle, P. *Chemosphere* **1993**, *40*, 457-512.

Chapter 23

Multiphase Equilibrium in Mixtures of [C$_4$mim][PF$_6$] with Supercritical Carbon Dioxide, Water, and Ethanol: Applications in Catalysis

Vesna Najdanovic-Visak, Ana Serbanovic, José M. S. S. Esperança,
Henrique J. R. Guedes, Luis P. N. Rebelo,
and Manuel Nunes da Ponte

Instituto de Tecnologia Química e Biológica, Universidade Nova de Lisboa,
Aptd. 127, 2781–901 Oeiras, Portugal
REQUIMTE, Deparment of Chemistry, Universidade Nova de Lisboa,
2829–516 Caparica, Portugal
*Corresponding author: mnponte@itqb.unl.pt

The ionic liquid [C$_4$mim][PF$_6$] and supercritical carbon dioxide produce multiphase systems when mixed with ethanol and water. Mixtures of these four solvents can be made to go, by small changes in composition, through a succession of phase changes, involving one, two and three-phase situations. Increasing carbon dioxide pressure induces first the appearance of an intermediate liquid phase and later the merging of this phase with the gas, leaving all the ionic liquid in a separate, denser liquid. This succession is suitable to carry out reaction cycles in ionic liquid-based solvents, with complete recovery of the reaction product by CO$_2$ decompression.

Introduction

Supercritical carbon dioxide (ScCO$_2$) is a well-known clean solvent that found many applications, especially in separation and fractionation of natural products (*1*). More recently, it is being extensively studied as a green solvent for chemical reaction processes (*2,3*).

The use of ScCO$_2$ + an ionic liquid as an ideal combination for many chemical processes was first suggested by Blanchard *et al.* (*4*). These authors reported that mixtures of ScCO$_2$ with the ionic liquid [C$_4$mim][PF$_6$] show gas-liquid equilibrium behaviour whereby carbon dioxide can dissolve significantly into the [C$_4$mim][PF$_6$]-rich liquid phase, but no ionic liquid dissolves in the gas phase. Blanchard et al. (*5*) have also shown that organic compounds can be extracted from [C$_4$mim][PF$_6$] using ScCO$_2$.

This work has led to the study of carbon dioxide as an extracting solvent for several reaction processes in ionic liquids, including its use in continuous flow mode, bringing in reactants and taking out products (*6*).

In a more recent development, Scurto et al. (*7*) have presented phase behaviour results of [C$_4$mim][PF$_6$] + methanol + CO$_2$. Their data show the formation of an additional liquid phase at relatively low pressures, leading to three phase, liquid-liquid-gas equilibrium. The same authors (*8*) have also found that the introduction of gaseous or liquid carbon dioxide into a mixture of water and an ionic liquid can cause the separation of both hydrophobic and hydrophilic ionic liquids from aqueous solution, with the formation of an intermediate third phase.

Almost fifty years ago, Francis (*9*) published an extensive account of phase behaviour of binary and ternary mixtures containing liquid carbon dioxide. He lists 21 systems where completely miscible liquids are separated into two liquid phases by the introduction of CO$_2$, and many others where partial immiscibility of two liquids is enhanced. The behaviour discovered by Scurto el al. is therefore an expression in ionic liquid systems of the long-known capability of carbon dioxide as an inducer of immiscibility.

The importance of this effect is that carbon dioxide may be used, at relatively low pressures, to control the number of phases in a reactive system where the reaction takes place in the ionic liquid.

There is another interesting feature of the phase behaviour of [C$_4$mim][PF$_6$] + methanol + CO$_2$. For the mixtures with lower content in ionic liquid, Scurto et al. (*7*) have shown that an increase in pressure leads to the disappearance of the intermediate liquid phase, by merging with the upper gas phase, through a critical point. The resulting fluid phase did not contain any ionic liquid, which was retained in the lower liquid phase.

Phase behaviour of carbon dioxide + water + an alcohol + an ionic liquid

Swatloski et al. (*10,11*) and Najdanovic-Visak et al. (*12*) discovered that the addition of water to binary mixtures of imidazolium-based ionic liquids with several alcohols, which show extensive immiscibility areas, increased mutual solubility, until a single phase was formed. Recent work from this laboratory (*13*) has shown that this surprisingly large co-solvent effect extends over wide ranges of temperatures and compositions. This work has also included the study of pressure and isotope-substitution effects on phase transitions.

These water + alcohol + ionic liquid mixtures, due to the variety of their phase transitions, are especially appropriate solvents to carry out reactions where the rate may be controlled by switching, for instance, from two phases to one phase, by small composition changes. The separation of products from the reaction mixture would, however, become more complicated than in pure biphasic reactions.

Najdanovic-Visak et al. (*14*) have recently shown that it is possible to separate [C$_4$mim][PF$_6$] from water/ethanol mixtures using CO_2. As carbon dioxide is added to [C$_4$mim][PF$_6$]/ethanol/water, a third phase starts to form between the liquid and gas phases, in a similar fashion to what Scurto et al. reported on [C$_4$mim][PF$_6$] + methanol. At higher pressures, critical points involving the intermediate liquid phase and the vapour were observed, for different water/ethanol molar ratios.

These findings indicate that the phase behaviour discovered by Scurto et al. is exhibited by many systems with ionic liquids, and that it is not confined to mixtures of low content in ionic liquid, as those authors suggested.

A thermodynamic analysis can be based on a succession of phase diagrams for ternary diagrams, as shown in Figure 1.

In the first (left) diagram of the Figure, at low pressures, simple liquid-vapour equilibrium is observed. The tie-line connecting the liquid and the vapour sides will depend on the initial composition of the liquid mixture A+B. The vapour side is essentially pure CO_2.

In the second diagram, a third phase L$_2$ has already been formed. An increase of CO_2 pressure provoked further dissolution of carbon dioxide in the L$_1$ phase and induced the separation of L$_2$. The separation begins at the pressure where the overall composition of the mixture (phases liquid and vapour combined) hits the black triangle in the diagram, which corresponds to the three-phase area. This depends on the initial composition of the liquid mixture A+B, on the amount of CO_2 added and on the available volume in the vessel containing the mixture (that is, the total volume less the volume occupied by the liquid phase). There is a limited range of compositions that lead to the formation of the third phase L$_2$, and its appearance (or not) is therefore dependent on the conditions of the experiment being carried out.

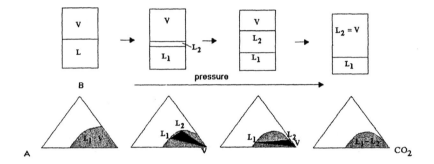

Figure 1 – Correspondence between the succession of states observed when carbon dioxide is added isothermally to a mixture of ionic liquid (A) and an alcohol or alcohol + water (B). (See page 3 of color insert.)

The third diagram on the left corresponds to the situation depicted above it, where phase L_2 has grown significantly in volume by further addition of carbon dioxide. The black triangle has moved, indicating that phase L_1 is now much richer in component A (the ionic liquid) and L_2 has increased its content of component B and carbon dioxide. In some cases of very precise matching of compositions, a four-phase equilibrium situation might appear at a fixed pressure. A four-phase line in the (p,T) projection of a phase equilibrium diagram of a ternary mixture is limited between two fixed (p,T) points, the lower and the upper critical end points (LCEP and UCEP), as is very clearly explained in the paper of Wendland et al. (*15*).

When carbon dioxide continues to be added to the system, two types of event may be observed: (a) either the pressure vessel fills up to the top with liquid L_2, or (b) a critical point $(L_2=V)$ is formed, and the two phases become indistinguishable. This last situation is the one depicted in Figure 1. It can only be observed, of course, when binary mixtures of components B and CO_2 exhibit vapour-liquid critical points. The important peculiarity of ionic liquid-containing mixtures is, as pointed out by Scurto et al., that the ionic liquid is totally confined to the L_1 phase when the critical point $L_2=V$ is reached.

This last observation allows further thermodynamic analysis of the phase diagram, in terms that are now specific of ionic liquid solutions, at pressures close to the critical pressure

In the case of the methanol–based mixtures studied by Scurto et al. (*7*), the $L_2 + V$ part of the pressure vessel (Figure 1) can be viewed as a binary CO_2 + methanol. As the authors point out, the critical pressures for all compositions where a critical point is observed are within experimental error of the critical pressure determined for the binary. Their case where no formation of L_2 was found, corresponding to a high concentration of IL (49.3% molar in the initial liquid mixture) can be interpreted as a situation where the small free volume above the liquid did not allow enough space for the required quantity of methanol to be drawn into the vapour phase, so that it would cross into the vapour-liquid equilibrium area. This is depicted in Figure 2, where two different paths are drawn for the vapour phase. The path on the left produces a second L_2 liquid phase when it meets the two-phase area envelope, and leaves this area at higher pressures through the critical point. The path on the right, which does not cross this envelope, should correspond to the situation where no third phase is formed. As this is dependent on the available volume in the cell, the conclusion of Scurto et al. that, "when the concentration in IL is high, apparently it is not possible to induce liquid-liquid phase split by the addition of CO_2" should be revisited.

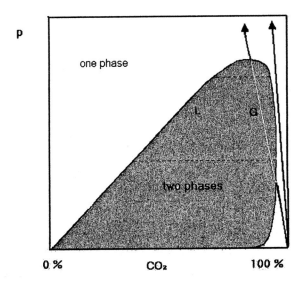

Figure 2 – Binary mixture isothermal phase diagram, where two different paths for the pressure increase on carbon dioxide addition are shown.

In the case presented by Najdanovic-Visak et al. (*14*), component B is a mixture of water and ethanol. The diagrams of Figure 1 do not apply rigorously to this situation, but the quaternary IL + CO_2 + water + ethanol may be treated as a pseudo-ternary. Especially in the vicinity of the critical points ($L_2 = V$) observed by these authors, the $L_2 + V$ part of the overall mixture can be interpreted in terms of the ternary phase diagram for CO_2 + water + ethanol.

As in this case there is one more degree of freedom than in the methanol + IL mixtures, there are more experiments where no critical point is observed, because the compositions are not the right ones, and the high pressure cell is filled with liquid L_2 (with disappearance of V) at pressures lower than the critical. When critical points were indeed observed, those authors concluded, by comparison with the critical points for the pure ternary mixture (without IL) that water is taken out preferentially from ethanol-rich IL mixtures, while for water-rich initial mixtures, ethanol is preferentially withdrawn from the ionic liquid. This effect might be related to the more recent finding by Najdanovic-Visak et al. (*13*) that pressure increases mutual solubility of $[C_4mim][PF_6]$ and ethanol, but decreases it in the case of the same ionic liquid and water.

Applications in catalysis

The cascade of phase changes in ionic liquid aqueous induced by ethanol and carbon dioxide can be used to allow reaction cycles to proceed as shown in Figure 3.

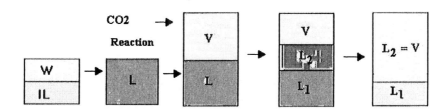

Figure 3 – Water + [C₄mim][PF₆] mixtures change the number of phases with successive addition of ethanol (first transition) and carbon dioxide. (See page 3 of color insert.)

A reaction usually carried out in biphasic water + IL conditions, can benefit from monophasic conditions, for increased rates, by addition of ethanol, without losing the advantages of biphasic systems for catalyst recycling and product separation, because carbon dioxide can then be used to extract the reaction products and regenerate the ionic liquid phase.

The reaction chosen (*14*) to carry out a proof-of-principle experiment was the (usually slow) epoxidation of isophorone by hydrogen peroxide, catalysed by sodium hydroxide:

isophorone isophorone oxide

Bortolini et al. (*16*) carried out this epoxidation, and the epoxidation of several other electrophilic alkenes, dissolved in [C₄mim][PF₆], by contact with an aqueous solutions of hydrogen peroxide. These reactions were performed in biphasic conditions, due to the above-mentioned immiscibility of water and the ionic liquid. Ethyl acetate was used to extract the products from the reaction mixture.

The difference in the rate of this reaction when carried out either in biphasic conditions, similar to those of Bortolini et al. *(16)* or in one phase, by addition of ethanol to the water + ionic liquid immiscible system, *(14)* is shown in Figure 4.

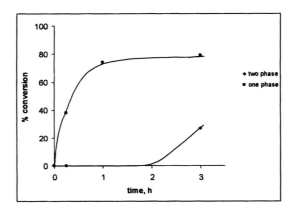

Figure 4 - Yield of the epoxidation reaction of isophorone carried out in a one-phase [C₄mim][PF₆] + water + ethanol solvent or in a biphasic mixture of [C₄mim][PF₆] and water (ref.14).

In the one-phase conditions, extraction with carbon dioxide, at 12 MPa and 313 K, completely removed the reaction product from the reaction mixture.

In a more recent work, Bortolini et al. *(17)* contact their two-phase reaction mixture (for a different substrate, 2-cyclohexen-1-one), with ScCO₂, also at 313 K, but at the somewhat more drastic pressure of 20 MPa. Their results are essentially solubilities of the epoxide in the carbon dioxide-rich phase. They conclude that extraction of the reaction product with carbon dioxide in those conditions is viable.

Acknowledgements

This work was financially supported by Fundação para a Ciência e a Tecnologia (FC&T, Portugal), through contracts POCTI/EQU/35437/99 and POCTI/QUI/38269/2001. VNV, AS and JMSSE thank FC&T for doctoral fellowships.

References

1. Brunner, G. Gas Extraction; Springer: New York, **1994**
2. Noyori R. (ed), "Supercritical Fluids: Introduction", *Chem. Rev.*, **1999**, *99 (2)*, 353-354.
3. Jessop, P.G.; Leitner, W.; (eds). Chemical Synthesis in Supercritical Fluids, Wiley-VCH, Weinheim, **1999**.
4. Blanchard, L.A.; Hancu, D.; Beckman, E.J.; Brennecke, J.F. *Nature*, **1999**, 399, 28
5. Blanchard, L.A.; Brennecke, J.F. *Ind. Eng. Chem. Res.*; **2001**, *40*, 287-292
6. Sellin,M.F.; Webb; P. B.; Cole-Hamilton, D.J. *Chem. Commun.*, **2001**, 781
7. Scurto, A.M.; Aki, S.N.V.K.; Brennecke, J.F. *J. Am. Chem. Soc.*, **2002**, *124*, 10276
8. Scurto, A.M.; Aki, S.N.V.K.; Brennecke, J.F. *Chem. Commun.*, **2003**, 572
9. Francis, A.W. J. Phys. Chem. **1954**, *58*, 1099
10. Swatloski, R.P.; Visser, A.E.; Reichert, W.M.; Broker, G.A.; Farina, L.M.; Holbrey, J.D.; Rogers, R.D. *Chem. Commun.*, **2001**, *20*, 2070-2071.
11. Swatloski, R.P.; Visser, A.E.; Reichert, W.M.; Broker, G.A.; Farina, L.M. ; Holbrey, J.D.; Rogers, R.D. *Green Chemistry*, **2002**, *4*, 81-87
12. Najdanovic-Visak, V.; Esperança, J.M.S.S.; Rebelo, L.P.N.; Nunes da Ponte, M.; Guedes, H.J.R.; Seddon, K.R.; Szydlowski, J. *Phys. Chem. Chem. Phys.*, **2002**, *4*, 1701-1703
13. Najdanovic-Visak, V.; Esperança, J.M.S.S.; Rebelo, L.P.N.; da Ponte, M.N.; Guedes, H.J.R.; Seddon, K.R.; Sousa, H.C.; Szydlowski, J. *J. Phys. Chem. B* **2003**, *107*, 12797-12807
14. Najdanovic-Visak, V.; Serbanovic, A.; Esperança, J.M.S.S.; Guedes, H.J.R.; Rebelo, L.P.N.; Nunes da Ponte; M. *ChemPhysChem*, **2003**, *4*, 520

15. Wendland, M.; Hasse, H.; Maurer, G. *J. Supercritical Fluids*, **1993**, *6*, 211-222
16. Bortolini, O.; Conte, V.; Chiappe, C.; Fantin, G.; Fogagnolo, M.; Maietti, S. *Green Chemistry*, **2002**, *4*, 94-96
17. Bortolini, O.; Campestrini, S.; Conte, V.; Fantin, G.; Fogagnolo, M.; Maietti, S. *Eur. J. Org. Chem.*, **2003**, *24*, 4804-4809

Author Index

Subject Index

A

Activity coefficients
  high boiling solutes in ionic liquid,
    188–189
  nonanal in mixture with 1-methyl-
    3-ethylimidazolium
    bis(trifluoromethylsulfonyl)imid
    e ([emim][ntf$_2$]), 190*f*
Alcohols. *See* 1-Butyl-3-
  methylimidazolium
  bis(trifluoromethylsulfonyl)imide
  ([bmim][NTf$_2$]); Multiphase
  equilibrium in mixtures
1-Alkyl-3-methylimidazolium cations
  modeled charges, 123*t*
  nomenclature, 137*f*
1-Alkyl-3-methylimidazolium
  chlorides
  solid-liquid equilibria in alcohols,
    257
  solubilities in octanol and water,
    258
  See also Phase equilibria
1-Alkyl-3-methylimidazolium family
  ab initio calculations of partial
    charges, 144
  ab initio calculations of torsion
    energy profiles, 139–141
  bond lengths and angles, 136–139
  charge distribution in cations by
    different force fields, 143*t*
  comparing experimental and
    predicted molar volumes, 285*f*
  crystalline phase, 145–146
  external force field
    parameterization, 141*t*
  intermolecular potential, 141–144
  ionic liquid solutions, 271
  liquid phase simulation, 146*t*, 147

molecular dynamics (MD)
  simulations, 145
molecular geometry, 136–139
nomenclature of cations, 137*f*
non-bonded interactions
  parameters, 142*t*
validation, 144–147
See also Gas interactions with ionic
  liquids
1-Alkyl-3-methylimidazolium salts.
  See Dioxouranate(VI) species in
  ionic liquids
All-Atom (AA) force field approach
  ab initio calculations of partial
    charges, 144
  ab initio calculations of torsion
    energy profiles, 139–141
  1-alkyl-3-methylimidazolium
    family, 135–144
  1-alkyl-3-methylimidazolium non–
    bonded interactions parameters,
    142*t*
  bond lengths and angles, 136–139
  charge distribution in ionic liquid
    cations, 143*t*
  comparison between experimental
    X-ray geometries and single-
    molecule *ab initio* calculations,
    138*t*
  external force field
    parameterization, 141*t*
  force field model, 135
  intermolecular potential, 141–144
  ionic liquid models, 136*t*
  ionic liquid simulations, 119
  molecular geometry, 136–139
  stretching and bending force-field
    parameters, 139*t*
  torsional force-field parameters,
    140*t*

RETURN TO: **CHEMISTRY LIBRARY**
100 Hildebrand Hall • 510-642-3753

LOAN PERIOD	1	2	3
4		5	6

ALL BOOKS MAY BE RECALLED AFTER 7 DAYS.
Renewals may be requested by phone or, using GLADIS,
type **inv** followed by your patron ID number.

DUE AS STAMPED BELOW.

NON-CIRCULATING
UNTIL: 8/18/05

DEC 20 2005

AUG 17

DEC 20

JUN 30
DEC 20